T0269259

Ground-Source Heat Pumps

Ground-Source Heat Pumps

Fundamentals, Experiments and Applications

Ioan Sarbu
Department of Building Services Engineering,
Polytechnic University of Timisoara, Romania

Calin Sebarchievici
Department of Building Services Engineering,
Polytechnic University of Timisoara, Romania

AMSTERDAM • BOSTON • HEIDELBERG • LONDON
NEW YORK • OXFORD • PARIS • SAN DIEGO
SAN FRANCISCO • SINGAPORE • SYDNEY • TOKYO
Academic Press is an imprint of Elsevier

Academic Press is an imprint of Elsevier
125, London Wall, EC2Y 5AS.
525 B Street, Suite 1800, San Diego, CA 92101-4495, USA
225 Wyman Street, Waltham, MA 02451, USA
The Boulevard, Langford Lane, Kidlington, Oxford OX5 1GB, UK

ISBN: 978-0-12-804220-5

British Library Cataloguing-in-Publication Data
A catalogue record for this book is available from the British Library

Library of Congress Cataloging-in-Publication Data
A catalog record for this book is available from the Library of Congress

For Information on all Academic Press publications
visit our website at http://store.elsevier.com/

Publisher: Joe Hayton
Acquisition Editor: Raquel Zanol
Editorial Project Manager: Mariana Kühl Leme
Editorial Project Manager Intern: Ana Claudia A. Garcia
Production Project Manager: Kiruthika Govindaraju
Marketing Manager: Louise Springthorpe
Cover Designer: Greg Harris

CONTENTS

AUTHOR BIOGRAPHIES

Ioan Sarbu is a professor and head of the Department of Building Services Engineering at the Polytechnic University of Timisoara, Romania. He obtained a diploma in civil engineering from the 'Traian Vuia' Polytechnic Institute of Timisoara in 1975 and a PhD degree in civil engineering from the Timisoara Technical University in 1993. He is a European Engineer, as designated by European Federation of National Engineering Associations (Brussels) in 2001.

His main research interests are related to refrigeration systems, heat pumps and solar energy conversion. He is also active in the field of thermal comfort and environmental quality, energy efficiency and energy savings, and numerical simulations and optimisations in building services. Additionally, he is a member of the American Society of Heating, Refrigerating and Air-Conditioning Engineers (ASHRAE), International Association for Hydro-Environment Engineering and Research (IAHR), Romanian Association of Building Services Engineers (RABSE), and Society for Computer-Aided Engineering (SCAE). He is a doctoral degree advisor in the civil engineering branch, an expert reviewer on the National Board of Scientific Research for Higher Education (Bucharest), vice president of the National Board of Certified Energetically Auditors Buildings (Bucharest), a member of the National Council for Validation of University Titles, Diplomas and Certificates, a member of the Technical Council for Civil Engineering from Ministry of Regional Development and Public Administration, and a reviewer of the *Journal of Thermal Science and Technology, Energy Conversion and Management, Applied Thermal Engineering, International Journal of Refrigeration, International Journal of Sustainable Energy, Energy and Buildings, and Energy Efficiency.*

Honours (citation, awards, etc.): Award, Romanian General Association of Engineers, 1997; Diploma, Outstanding Scientist of the 21st Century, IBC, 2005; World Medal of Freedom, ABI, 2006; Plaque, Gold Medal for Romania, ABI, 2007; Lifetime Achievement Award, ABI, 2008; Ultimate Achiever Award Certificate for Engineering, IBC, 2009; Plaque, Hall of Fame for Distinguished Accomplishments in Science and Education, ABI, 2009; The Albert Einstein Award of Excellence, ABI, 2010; Diploma, Top 100 Engineers, IBC, 2012; Cambridge Certificate for Outstanding Engineering Achievement, IBC, 2013; Medal of Achievement, IBC, 2014; Listed in several *Who's Who* publications (e.g. *Who's Who in the World, Who's Who in Science and Engineering,* and *Who's Who in America*) and other biographical dictionaries.

He has published 12 books, 4 book chapters, more than 120 articles in indexed journals and about 50 articles in proceedings of international conferences. He is also the author of 5 patent certificates and of up to 20 computer programmes.

Calin Sebarchievici is a lecturer in the Building Services Engineering Department at the Polytechnic University of Timisoara, Romania. He obtained a diploma in building services engineering from the Polytechnic University of Timisoara in 2003. He received his master's degrees from the same university and Bologna University, Italy, in 2004 and 2007, respectively, and his PhD degree in civil engineering from the Polytechnic University of Timisoara in 2013.

His research is focused on air conditioning, heat pump and refrigeration systems. He is also active in the field of thermal comfort, energy efficiency and energy savings. Additionally, he is a member of the ASHRAE, RABSE and GEOEXCHANGE Romanian Society. He is co-author of 1 book, 2 book chapter, 15 journal articles and more than 10 conference proceeding publications.

PREFACE

Energy security is the ability of a nation to deliver the energy resources needed to ensure its welfare and implies a secure supply and stable prices. Energy is vital for the progress and development of a nation's economy. The economic growth and technological advancement of every country depends on energy, and the amount of available energy affects that country's quality of life. Economy, population and per capita energy consumption have increased the demand for energy during the last several decades. Fossil fuels continue to supply much of the energy used worldwide, and oil remains the primary energy source. However, fossil fuels are a major contributor to global warming. The awareness of global warming has intensified in recent times and has reinvigorated the search for energy sources that are independent of fossil fuels and contribute less to global warming.

Because of the reduction of the world's fossil fuel reserves and strict environmental protection standards, one main research direction in the construction field has become the reduction of energy consumption, including materials, technology and building plans with lower specific energy needs, along with high-performance equipment.

Renewable energy sources have significant potential to contribute to sustainable development. The Renewable Energy Directive 2009/28/EC opened up a major opportunity for further use of heat pumps for both the heating and cooling of new and existing buildings and domestic hot-water (DHW) production. Heat pumps (HPs) are used in an alternative heating/cooling system that is more energy efficient and less polluting than a traditional system (liquid or gas fuel boiler). A large number of ground-source heat pump (GSHP) systems have been used in residential and commercial buildings throughout the world due to the attractive advantages of high energy and environmental performance.

This book treats a modern issue of great current interest at a high scientific and technical level, based both on original research and achievements and on the synthesis of consistent bibliographic material to meet the increasing need for modernisation and for greater energy efficiency of building services to significantly reduce CO_2 emissions. The book mainly presents a detailed theoretical study and experimental investigations on GSHP technology, concentrating on ground-coupled heat pump (GCHP) systems. The text offers a comprehensive and consistent overview of geothermal HP applications, performance and combination with heating/cooling systems and covers the technical, economic and energy savings aspects related to the design, modelling and operation of these systems. This book is structured as seven chapters. Chapter 1 summarises a description of renewable energy, concentrating on geothermal energy, and presents the operation principle of an HP and the necessity for using HPs in the heating/cooling systems of buildings.

Chapter 2 discusses vapour compression-based heat pump (VCBHP) systems and describes the theoretical and real thermodynamic cycles and their calculation, as well as the operation regimes of a VCBHP with an electro-compressor. Technical aspects such as the partial load performances and the calculation of the greenhouse gas emissions of heat pumps are considered. Finally, this chapter presents the energy, economic and environmental performance criteria that show the opportunity to implement an HP in a heating/cooling system.

Chapter 3 presents a study on the recent development of possible substitutes for non-ecological refrigerants for heating, ventilating, air conditioning and refrigeration equipment based on thermodynamic, physical and environmental properties and total equivalent warming impact analysis. This study contains information regarding the environmental pollution produced by the working fluids of HPs, air conditioning, and commercial refrigeration applications and the ecological refrigerant trend. Overall, the study is useful for those readers who are interested in the current status of alternative refrigerant development related to vapour compression-based refrigeration systems. The study describes the selection of refrigerants adapted to each utilisation based on its thermodynamic, physical and environmental properties and technological behaviour, as well as the use of constraints as the principal aspect of environmental protection. This chapter also explores the studies reported with new refrigerants in HPs, domestic and commercial refrigerators, chillers and air conditioners.

Chapter 4 presents a detailed description of the refrigeration compressor types (reciprocating, rotary screw, centrifugal and scroll compressors) and the HP types. Important information on the selection of the heat source and HP systems and DHW production for nearly zero-energy buildings are discussed. In addition, some installation instructions for HPs and examples of HP utilisation are presented. Finally, the renewable energy source contribution from HP sales in the European Union is shown.

Chapter 5 presents a detailed description of ground characteristics, GSHPs and GSHP development. It also discusses the most common simulation models and programs of vertical ground heat exchangers (GHEs) and borehole heat exchangers (BHEs) currently available, and describes different applications of the models and programs (simulation of ground thermo-physical capacity, a one-dimensional transient BHE model, modelling the interactions between ground temperature variations and performance of GCHPs, and a vertical GHE design based on hourly load simulations). Additionally, a new groundwater heat pump using a heat exchanger with a special construction, which was tested in a laboratory with the possibility of obtaining better energy efficiency with combined heating and cooling by GCHP, are presented. Finally, the advanced engineering applications of hybrid GCHP systems and environmental performance are also briefly analysed.

Chapter 6 performs an energy-economic analysis and compares different heating systems in terms of energy consumption, thermal comfort and environmental impact. A computational model of annual energy consumption is developed for an air-to-water HP based on the degree-day method and the bin method implemented by a computer program. A comparative economic analysis of heating solutions for a building is performed from a case study, and the energy and economic advantages of building heating solutions with a water-to-water HP are reported. The energy, economic and environmental performance of a closed-loop GCHP system is also analysed. In addition, the main performance parameters (energy efficiency and CO_2 emissions) of radiators and radiant floor heating systems connected to GCHPs are compared. These performances were obtained from site measurements in an office. Furthermore, the thermal comfort for these systems is compared using the ASHRAE Thermal Comfort program, and a mathematical model for numerical modelling of the thermal emission at radiant floors is developed and experimentally validated. Additionally, two numerical simulation models of useful thermal energy and the system coefficient of performance (COP_{sys}) in heating mode are developed using the Transient Systems Simulation (TRNSYS) software. The simulations obtained from the TRNSYS software are analysed and compared to experimental measurements. Finally, important information for control of HP heating and cooling systems is included.

Chapter 7 focuses on the energy and environmental analysis and modelling of a reversible GCHP. One of the main innovative contributions of this study is in the achievement and implementation of an energy-operational optimisation device for the GCHP system using quantitative adjustment with a buffer tank and a variable speed circulating pump. Experimental measurements are used to test the performance of the GCHP system at different operating modes. The main performance parameters (energy efficiency and CO_2 emissions) are obtained for 1 month of operation using both classical and optimised adjustments of the GCHP system. A comparative analysis of these performances for both heating and cooling and DHW with different operation modes is performed. Additionally, two simulation models of thermal energy consumption in heating/cooling and DHW operation are developed using TRNSYS software. The simulations obtained using TRNSYS software are analysed and compared to the experimental measurements. The second objective of this chapter is to present the COP of a horizontal GCHP system and the temperature distributions measured in the ground heating season. Finally, the use of a numerical model of heat transfer in the ground for determining the temperature distribution in the vicinity of pipes is described.

This book provides a source of material for all of those involved in the field, whether as a student, scientific researcher, industrialist, consultant or government agency with responsibility in this area.

CHAPTER 1

Introduction

1.1 GENERALITIES

Buildings are an important part of European culture and heritage, and they play an important role in the energy policy of Europe. An economic strategy of sustainable development certainly promotes efficiency and rational energy use in buildings, which are the major energy consumer in Romania and the other member states of the European Union (EU). EU energy consumption patterns reveal that buildings are the greatest energy consumer, using approximately 40% of the total energy demand, followed by industry and transportation, which consume approximately 30% each [1]. Buildings offer the greatest and most cost-effective potential for energy savings. Studies have also shown that saving energy is the most cost-effective method for reducing greenhouse gas (GHG) emissions.

Currently, heating is responsible for almost 80% of the energy demand in houses and utility buildings, used for the purpose of space heating and hot-water generation, whereas the energy demand for cooling is growing yearly. Pollution emissions of GHGs, rising energy demand and an increasing dependence on imports are important energy problems for any country to solve.

To realise the ambitious goals for reducing the consumption of fossil fuel as primary energy and the related CO_2 emissions, and to reach the targets of the Kyoto Protocol, improved energy efficiency and the use of renewable energy in the existing building stock must be addressed in the near future.

The European strategy to decrease energy dependence rests on two objectives: the diversification of the various sources of supply and policies to control consumption. The key to diversification is renewable energy sources (RES) because they have significant potential to contribute to sustainable development [2].

This chapter summarises a description of renewable energy, concentrating on geothermal energy, and presents the operation principle of a heat pump (HP) and the necessity for using HPs in the heating/cooling systems of buildings.

1.2 RENEWABLE ENERGY

The term "renewable energy" refers to energy that is produced from natural resources that have the characteristics of inexhaustibility over time and natural renewability.

RES include wind, solar, geothermal, biomass and hydro energies [3]. An efficient utilisation of renewable resources has significant potential to both stimulate the economy and reduce pollution. Thus, many governments have implemented various policies that support renewable energy generation. One of the key components of any renewable energy policy is the setting of renewable energy targets.

Ground-Source Heat Pumps. DOI: http://dx.doi.org/10.1016/B978-0-12-804220-5.00001-1

Numerous efforts have been undertaken by developed countries to implement different renewable energy technologies, and in particular, the use of wind energy has increased over the last few years [4]. For example, the Netherlands, Germany, India and Malaysia use wind turbines to produce electricity [5]. In northwestern Iran, mineral materials are used for the production of geothermal energy, and in Iceland, 70% of factories use geothermal energy for industrial purposes [6].

Although Romania has a high potential of RES, in 2010, the RES share of the total energy consumption was 23.4%. Romania ranked second place in the EU for the portion of energy from renewable sources out of gross final consumption from 2006 to 2010 [7].

Among the energy alternatives to fossil fuels, RES such as solar and wind are more available.

On 23 April 2009, the European Parliament and the Council adopted the Renewable Energy Directive 2009/28/EC, which establishes a common framework for the promotion of RES. This directive has stated that by the year 2020, the average energy percentage from renewable sources will be approximately 20% of the total energy consumption (Table 1.1). The 2009/28/EC directive opens up a major opportunity for the further use of HPs for heating and cooling of new and existing buildings. HPs enable the use of ambient heat at useful temperature levels and need electricity or other form of energy to function.

Table 1.1 Renewable Energy Percent out of Total Energy Consumption for 2020

No.	Country	Code	Percentage (%)
1	Belgium	BE	13
2	Bulgaria	BG	16
3	Czech Republic	CZ	13
4	Denmark	DK	30
5	Germany	DE	18
6	Estonia	ES	25
7	Ireland	IE	16
8	Greece	GR	18
9	Spain	ES	20
10	France	FR	23
11	Italy	IT	17
12	Cyprus	CY	13
13	Latvia	LV	42
14	Lithuania	LT	23
15	Luxembourg	LU	11
16	Hungary	HU	13
17	Malta	MT	10
18	Netherlands	NL	14

(*Continued*)

Table 1.1 (Continued)

No.	Country	Code	Percentage (%)
19	Austria	AT	34
20	Poland	PL	15
21	Portugal	PT	31
22	Romania	RO	24
23	Slovenia	SI	25
24	Slovakia	SK	14
25	Finland	FI	38
26	Sweden	SE	49
27	United Kingdom	UK	15

The amount of ambient energy E_{res} captured by HPs to be considered as renewable energy shall be calculated in accordance with the following equation [8]:

$$E_{res} = E_U\left(1 - \frac{1}{SPF}\right) \tag{1.1}$$

where E_U is the estimated total usable thermal energy delivered by HPs and SPF is the estimated seasonal performance factor for these HPs.

Only HPs for which $SPF > 1.15/\eta$ shall be taken into account, where η is the ratio between the total gross production of electricity and the primary energy consumption for electricity production. For EU countries, the average η is 0.4, meaning that the minimum value of the seasonal performance factor should be 2.875.

1.3 GEOTHERMAL ENERGY

Geothermal energy is the energy stored in the Earth's crust. The temperature and pressure increase with increasing depth into the Earth's crust, and geothermal energy can be used more efficiently. Thus, the potential of geothermal energy is huge, but only a part of it can be used. Geothermal resources exist in low-enthalpy forms (corresponding to temperatures less than 200 °C), used mainly for direct heating applications; and high-enthalpy forms (corresponding to temperatures higher than 200 °C), suitable for electricity generation. The ground temperature can be considered relatively constant throughout the year, even starting from the lower depths. Romania, together with other neighbouring countries (Hungary and Serbia), has important low-enthalpy geothermal resources suitable for direct heating applications.

Ground-source heat pump (GSHP) systems use the ground as a heat source and sink to provide space heating and cooling as well as domestic hot water. GSHP technology can offer higher energy efficiency for air conditioning (A/C) compared to conventional A/C systems because the underground environment provides higher temperatures for heating and lower temperatures for cooling and experiences less temperature fluctuation than ambient air.

The first known record of the concept of using the ground as a heat source for a HP was found in a Swiss patent issued in 1912 [9]. Thus, the research on GSHP systems has been undertaken for more than a century. The first surge of interest in GSHP technology began in both North America and Europe after World War II and lasted until the early 1950s, when gas and oil became widely used as heating fuels. At that time, the basic analytical theory for the heat conduction of the GSHP system was proposed by Ingersoll and Plass [10], which served as a basis for the development of some of the later design programs.

The next period of intense activity on GSHPs started in North America and Europe after the first oil crisis of the 1970s, with an emphasis on experimental investigation. In the two following decades, considerable efforts were made to establish the installation standard and to develop design methods for vertical borehole systems [11–13].

To date, GSHP systems have been widely used in both residential and commercial buildings. It is estimated that GSHP system installations have grown continuously on a global basis, ranging from 10% to 30% annually in recent years [14].

This book presents a detailed theoretical study and experimental investigations on GSHP technology, concentrating on ground-coupled heat pump systems.

1.4 OPERATION PRINCIPLE OF A HP

A HP is a thermal installation that is based on a reverse Carnot thermodynamic cycle, which consumes drive energy and produces a thermal effect.

Any HP moves (pumps) heat E_S from a source with low temperature t_s to a source with a high temperature t_u, consuming drive energy E_D.

- A *heat source* can be:
  - a gas or air (outdoor air, warm air from ventilation, hot gases from industrial processes)
  - a liquid called 'generic water': surface water (river, lake or sea), groundwater, or discharged hot water (domestic, technologic or recirculated in cooling towers); or
  - ground, with the advantage of accessibility.
- *Heat consumer*. The HP yields thermal energy at a higher temperature, depending on the application of the heat consumer. This energy can be used for:
  - space heating, which is related to low-temperature heating systems: radiant panels (floor, wall, ceiling or floor-ceiling), warm air or convective systems; or
  - water heating (pools, domestic or technologic hot water).

  The heat consumer is recommended to be associated with a cold consumer. This can be performed with either a reversible (heating–cooling) or a double-effect system. In cooling mode, an HP operates exactly like central A/C.
- *Drive energy*. HPs can be used to drive different energy forms:
  - electrical energy (electrocompressor)
  - mechanical energy (mechanical compression with expansion turbines)
  - thermomechanical energy (steam ejector system)
  - thermal energy (absorption cycle); or
  - thermoelectrical energy (the Peltier effect).

REFERENCES

[1] Anisimova N. The capability to reduce primary energy demand in EU housing. Energy Build 2011;43:2747–51.
[2] Zamfir AL. Management of renewable energy and regional development: European experiences and steps forward. Theor Empir Res Urban Manage 2011;6(3):35–42.
[3] Hassan HZ, Mohamad AA. A review on solar-powered closed physisorption cooling systems. Renewable Sustainable Energy Rev 2012;16:2516–38.
[4] Ullah KR, Saidur R, Ping HW, Akikur RK, Shuvo NH. A review of solar thermal refrigeration and cooling methods. Renewable Sustainable Energy Rev 2013;24:490–513.
[5] Lucas T, Raoult-Wack AL. Immersion chilling and freezing in aqueous refrigerating media: review and future trends. Int J Refrig 1998;21:419–29.
[6] Abu Hamdeh NH, Al-Muhtaseb MA. Optimization of solar adsorption refrigeration system using experimental and statistical techniques. Energy Convers Manage 2010;511610–15.
[7] Colesca SE, Ciocoiu CN. An overview of the Romanian renewable energy sector. Renewable Sustainable Energy Rev 2013;24:149–58.
[8] Seppänen O. European parliament adopted the directive on the use of renewable energy sources. Rehva J 2009;46(1):12–14.
[9] Ball DA, Fischer RD, Hodgett DL. Design methods for ground-source heat pumps. ASHRAE Trans 1983;89(28):416–40.
[10] Ingersoll LR, Plass HJ. Theory of the ground pipe source for the heat pump. ASHVE Trans 1948;54:339–48.
[11] Kavanaugh SP, Rafferty K. Ground-source heat pumps, design of geothermal systems for commercial and institutional buildings. Atlanta, GA: ASHRAE; 1997.
[12] Bose JE, Parker JD, McQuiston FC. Design/data manual for closed-loop ground-coupled heat pump systems. Oklahoma State University for ASHRAE; 1985.
[13] Eskilson P. Thermal analysis of heat extraction boreholes [Doctoral thesis]. Sweden: University of Lund; 1987.
[14] Bose JE, Smith MD, Spitler JD. Advances in ground source heat pump systems – an international overview. In: Proceedings of the seventh international conference on energy agency heat pump. Beijing, China; 2002. p. 313–24.

CHAPTER 2

Vapour Compression-Based Heat Pump Systems

2.1 GENERALITIES

Ground-source heat pumps (GSHPs) are those with electro-compressor. The process of elevating low temperature heat to more than 38 °C and transferring it indoors involves a cycle of evaporation, compression, condensation and expansion (Figure 2.1). A non-chlorofluorocarbon refrigerant is used as the heat-transfer medium, which circulates within the heat pump (HP).

This chapter discusses the vapour compression-based heat pump (VCBHP) systems and describes the theoretical and real thermodynamic cycles and their calculation, as well as the operation regimes of a vapour compression HP with electro-compressor. Technical aspects such as the partial load performances and the calculation of the greenhouse gas (GHG) emissions of HPs are considered. Finally, this chapter presents the energy, economic and environmental performance criteria that allow for implementing an HP in a heating/cooling system.

2.2 THERMODYNAMIC CYCLE

2.2.1 Theoretical Cycle

The basic vapour-compression cycle is considered as one with isentropic compression, with no superheat of vapour and no subcooling of liquid (Figure 2.2).

Operational processes are outlined next:

$1-2$: isentropic compression in the compressor K, which leads to increased pressure and temperature from the values corresponding for evaporation p_0, t_0 to those of the condensation p_c, $t_2 > t_c$

$2-2'$: isobar cooling in the condenser C at pressure p_c from the temperature t_2 to $t_{2'} = t_c$

$2'-3$: isotherm–isobar condensation in the condenser C at pressure p_c and temperature t_c

$3-4$: isenthalpic lamination in expansion valve EV, leading the refrigerant from 3 state of the liquid at p_c, t_c in 4 state of wet vapour at p_0, t_0

$4-1$: isotherm–isobar evaporation in the evaporator E at pressure p_0 and temperature t_0.

In a theoretical vapour-compression cycle, the refrigerant enters the compressor at state 1, as saturated vapour, and is compressed isentropically to the condensation pressure. The refrigerant temperature increases during this isentropic compression process to well above the temperature of the surrounding medium. The refrigerant then enters

Ground-Source Heat Pumps. DOI: http://dx.doi.org/10.1016/B978-0-12-804220-5.00002-3

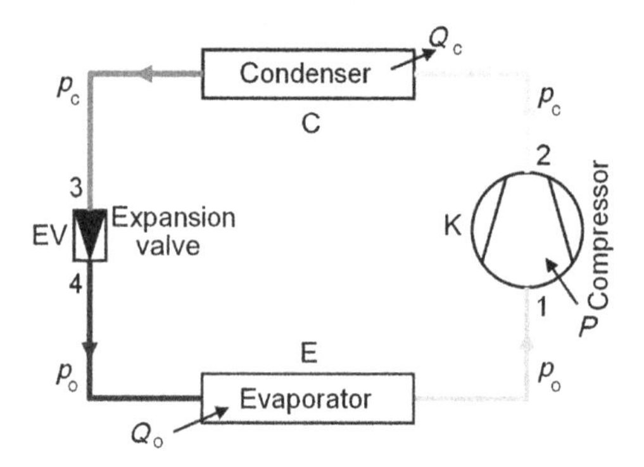

Figure 2.1 Schematic of a single-stage compression refrigeration system.

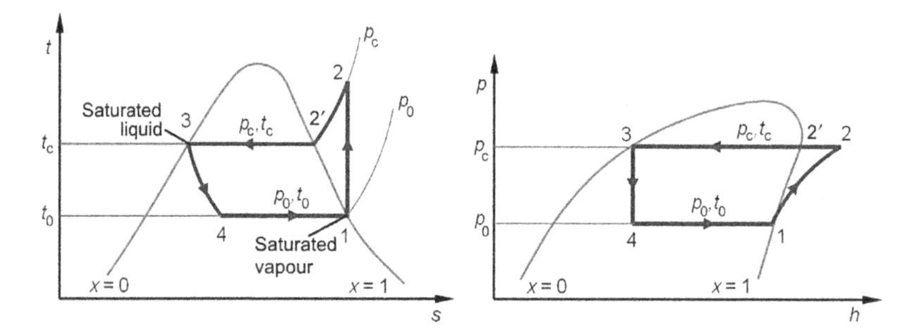

Figure 2.2 Single-stage vapour-compression process in t–s *and* p–h *diagrams.*

the condenser as superheated vapour at state 2 and leaves as saturated liquid at state 3 as a result of heat rejection to the surroundings. The refrigerant temperature at this state is still greater than the temperature of the surroundings. The saturated liquid refrigerant at state 3 is throttled to the evaporation pressure by passing it through an expansion valve. The refrigerant temperature drops below the temperature of the cold environment during this process. The refrigerant enters the evaporator at state 4 as a low-quality saturated mixture, and completely evaporates by absorbing heat from the cold environment. The refrigerant then leaves the evaporator as saturated vapour and re-enters the compressor, completing the cycle [1].

The specific compression work w, in kJ/kg, the specific cooling power q_0, in kJ/kg, the specific heat load at condensation q_c, in kJ/kg, volumetric refrigerating capacity q_{0v}, in kJ/m^3, the coefficient of performance COP are calculated for the previously presented processes as follows:

$$w = h_2 - h_1 \tag{2.1}$$

$$q_0 = h_1 - h_4 = h_1 - h_3 \tag{2.2}$$

$$q_c = h_2 - h_3 \tag{2.3}$$

$$q_{0v} = \frac{q_0}{v_1} = q_0 \rho_1 \tag{2.4}$$

$$\mathrm{COP} = \frac{q_c}{w} = \frac{h_2 - h_3}{h_2 - h_1} \tag{2.5}$$

Thermal power (capacity) of heat pump Q_{HP}, in kW, is expressed as:

$$Q_{HP} = mq_c \tag{2.6}$$

The power necessary for the isentropic compression P_{is}, in kW, may be calculated using the equation:

$$P_{is} = mw \tag{2.7}$$

The effective power P_{ef} on the compressor shaft is larger and is defined as:

$$P_{ef} = \frac{P_{is}}{\eta_{is}} \tag{2.8}$$

where η_{is} is the isentropic efficiency.

2.2.2 Real Cycle

The real operational processes (Figure 2.3) of a VBCHP deviate from the component processes of the theoretical cycle in the following ways:

- the compression process 1–2 in the compressor is adiabatic, but irreversible
- the heat exchange from the evaporator and the condenser is realised with finite temperature differences, imprinting on these processes an irreversible mark; the average temperature of the cold source t_s is higher than the evaporation temperature t_0, with the difference Δt_0, and the average temperature of the heat source t_u is lower that the condensation temperature t_c, with the difference Δt_c.
- the refrigerant flow through the system experiences pressure losses; and
- the equipment and pipes which the working fluid runs through exchange heat with the environment.

The irreversibility of the compression process increases the specific compression work to w' and increases the specific thermal load at condensation by Δq_c.

To assess the deviation in the degree of compression process 1–2 versus 1–2″, the (adiabatic) efficiency η_i of the compressor is defined as:

$$\eta_i = \frac{w}{w'} = \frac{w' - \Delta q_c}{w'} = 1 - \frac{\Delta q_c}{w'} < 1 \tag{2.9}$$

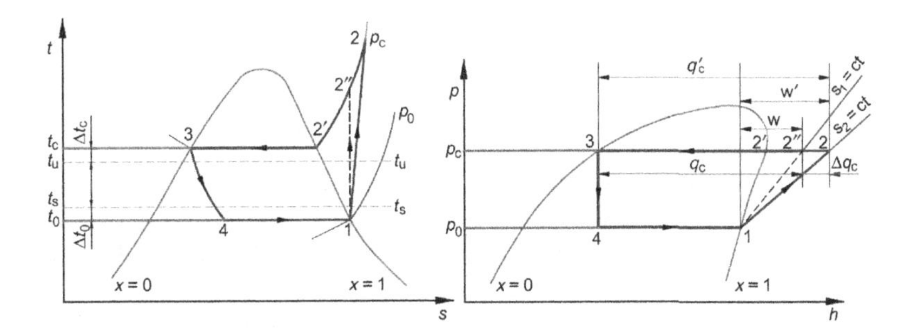

Figure 2.3 Real operation process in t–s *and* p–h *diagrams.*

where w is the specific isentropic compression work; w' is the specific adiabatic irreversible compression work; and Δq_c is the thermal load increase due to the irreversible compression process 1–2, given the isentropic process.

It follows that the real thermal efficiency $\mathrm{COP_r}$ (ε_r) of the cycle is given by:

$$\mathrm{COP_r} = \frac{q'_c}{w'} = \frac{q_c + \Delta q_c}{w + \Delta q_c} = \frac{q_c + \left(\frac{1}{\eta_i} - 1\right)w}{w + \left(\frac{1}{\eta_i} - 1\right)w} = (\mathrm{COP} - 1)\eta_i + 1 < \mathrm{COP} \tag{2.10}$$

where q'_c is the specific heat load at condensation in the real cycle with irreversible adiabatic compression.

The effective value of the thermal efficiency $\mathrm{COP_e}$ for the HP is diminished because of the mechanical and electrical losses of the electric motor-compressor assembly:

$$\mathrm{COP_e} = \mathrm{COP_r}\eta_{em} = [(\mathrm{COP} - 1)\eta_i + 1]\eta_{em} \tag{2.11}$$

where η_{em} is the electromechanical efficiency of compressor-motor assembly.

2.3 OPERATION REGIMES OF A HP

The operation regime of an HP is adapted to the existing heating system of a building. If the supply temperature is higher than the maximum supply temperature of the HP (55 °C), then the HP will only operate in addition to traditional sources of heat. In new buildings, a distribution system should be selected with a maximum supply temperature of 35 °C.

The following operating regimes are described next:

- *Monovalent regime.* For the univalent regime, the HP system meets the entire heat demand of the building at all times. The distribution system should be designed for a supply temperature below the maximum supply temperature of the HP. This operation regime is well suited for applications with supply temperatures of up to 65 °C. Systems with ground water or ground heat source collectors are operated as monovalent systems.
- *Bivalent regime.* A bivalent heating plant has two sources of heat. A HP with electrical action is combined with at least one heat source for solid, liquid or gaseous fuels. This regime can be bivalent-parallel (HP operates simultaneously with another heat source) or bivalent-alternate (usage of either the HP system or the other heat source).
  - Bivalent-parallel. The HP heats independently to a certain set point, at which an auxiliary heating system (electric element or boiler) is turned on and the two systems operate in tandem to meet the heating demand for a maximum supply temperature of up to 65 °C. This operation is used mainly with new air source systems or in renovations of old buildings.
  - Bivalent-alternate. The HP heats independently to a certain set point. Once this point is reached, a boiler meets the full heating demand. This operation is suitable for supply temperatures of up to 90 °C and is typically installed in renovated buildings.
- *Mono-energetic regime* is a bivalent operation regime in which the second heat source (auxiliary source) functions with the same type of energy (electricity) as the HP.

To make economic operation of a heating system with HPs possible, in some countries the electricity supplier provides special electricity tariffs for HPs. These prices usually assume that the electricity supply for HPs can be interrupted when the network is overloaded. For example, electricity supply for HP systems with the univalent operation regime may be discontinued three times in 24 h for more than 2 h. The operating time between two interruptions should not be less than the previous interruption. In the case of HP systems with bivalent operation, the electricity supply may be interrupted during the heating period for up to 960 h.

For existing buildings, the bivalent operation regime is recommended because a heat source exists, which can usually be used to further cover the peak loads of cold winter days with required supply temperatures of over 55 °C.

For new buildings, the univalent operation regime has proven useful because it may be interrupted. The HP can cover the annual heat demand, and the periods of interruption do not lead to disturbances in operation because, for example, floor heating interruption may not cause changes in the comfort temperature.

2.4 PERFORMANCES AND CO_2 EMISSION OF HP

The opportunity to implement an HP in a heating/cooling system is based on energy indicators and economic analysis.

2.4.1 Energy Efficiency

2.4.1.1 Coefficient of Performance

The operation of an HP is characterised by the coefficient of performance (COP) defined as the ratio between useful thermal energy E_t and electrical energy consumption E_{el}:

$$\mathrm{COP} = \frac{E_t}{E_{el}} \tag{2.12}$$

Seasonal coefficient of performance ($\mathrm{COP}_{seasonal}$) or average COP over a heating (cooling) season, often indicated as the seasonal performance factor (SPF) or annual efficiency, is obtained if in Eqn (2.12) is used summation of both usable energy and consumed energy during a season (year).

In the heating operate mode the HP COP is defined by the following equation:

$$\mathrm{COP}_{hp} = \frac{Q_{HP}}{P_e} \tag{2.13}$$

in which: Q_{HP} is the thermal power of HP, in W; P_e is the electric power consumed by the compressor of HP, in W.

In the cooling mode, an HP operates exactly like a central air-conditioner (A/C). The energy efficiency ratio (EER) is analogous to the COP but describes the cooling performance. The EER_{hp}, in Btu/(Wh) is defined as

$$\mathrm{EER}_{hp} = \frac{Q_0}{P_e} \tag{2.14}$$

in which: Q_0 is the cooling power of an HP, in British Thermal Unit per hour (Btu/h); P_e is the compressor power, in W.

The coefficient of performance of an HP in cooling mode is obtained by the following equation:

$$\mathrm{COP_{hp}} = \frac{\mathrm{EER_{hp}}}{3.412} \tag{2.15}$$

where value 3.412 is the transformation factor from Watt in Btu/h.

Figure 2.4 illustrates the COP variation of HPs in the heating operation mode, according to the source temperature t_s and the temperature at the consumer t_u [2].

The GSHP systems intended for ground-water or oven-system applications have heating COP ratings ranging from 3.0 to 4.0, and cooling EER ratings between 11.0 and 17.0. Those systems intended for closed-loop applications have COP ratings between 2.5 and 4.0 and EER ratings ranging from 10.5 to 20.0 [3]. The characteristic values of the SPF of modern GSHPs are commonly assumed to be approximately 4, meaning that four units of heat are gained per unit of consumed electricity.

The sizing factor (SF) of the HP is defined as the ratio of the HP capacity Q_{HP} to the maximum heating demand $Q_{\max}$:

$$\mathrm{SF} = \frac{Q_{\mathrm{HP}}}{Q_{\max}} \tag{2.16}$$

The SF can be optimised in terms of energy and economics, depending on the source temperature and its adjustment schedule.

From the energy balance of an HP:

$$E_{\mathrm{U}} = E_{\mathrm{S}} + E_{\mathrm{D}} \tag{2.17}$$

can highlight the link between the efficiency of a plant working as an HP ($\mathrm{COP_{hp}}$) and as a refrigeration plant ($\mathrm{COP_{rp}}$):

$$\mathrm{COP_{hp}} = \frac{E_{\mathrm{S}} + E_{\mathrm{D}}}{E_{\mathrm{D}}} = 1 + \frac{E_{\mathrm{S}}}{E_{\mathrm{D}}} = 1 + \mathrm{COP_{rp}} \tag{2.18}$$

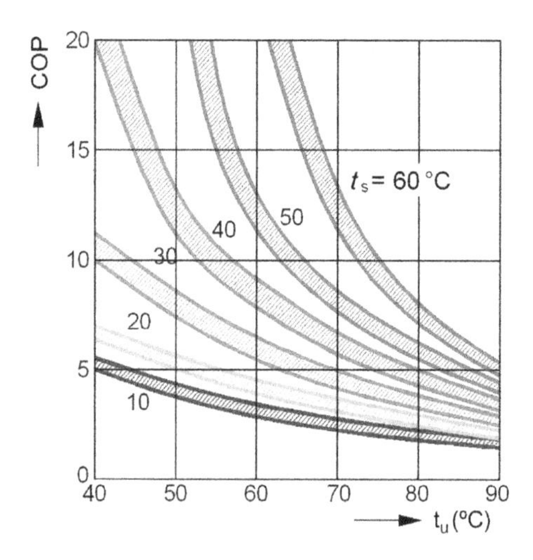

Figure 2.4 Efficiency variation of heat pumps.

The most effective systems are those that simultaneously use the produced heat and the adjacent refrigeration effect. The total energy efficiency of these systems is defined as:

$$\mathrm{COP}_{\mathrm{hp+rp}} = \frac{E_{\mathrm{U}} + E_{\mathrm{S}}}{E_{\mathrm{D}}} = \frac{E_{\mathrm{S}} + E_{\mathrm{D}} + E_{\mathrm{S}}}{E_{\mathrm{D}}} = \mathrm{COP}_{\mathrm{hp}} + \mathrm{COP}_{\mathrm{rp}} \tag{2.19}$$

The real HP efficiency ($\mathrm{COP}_{\mathrm{hp,r}}$) takes into account the energy losses that accompany both the accumulation and release of heat from the real processes Π_j. This can be evaluated using the following equation [4]:

$$\mathrm{COP}_{\mathrm{hp,r}} = \frac{t_{\mathrm{c}}}{t_{\mathrm{c}} - t_0}(1 - \Sigma\Pi_j) \tag{2.20}$$

where t_{c} and t_0 are the condensation and evaporation absolute temperatures of the refrigerant, in K.

The real efficiency of a HP with an electro-compressor can be determined from the following equation [5]:

$$\mathrm{COP}_{\mathrm{hp},r} = \frac{t_{\mathrm{u}} + \Delta t_{\mathrm{c}}}{t_{\mathrm{u}} + \Delta t_{\mathrm{c}} - (t_{\mathrm{s}} - \Delta t_{\mathrm{o}})}\eta_{\mathrm{r}}\eta_{\mathrm{i}}\eta_{\mathrm{m}}\eta_{\mathrm{em}} + \eta_{\mathrm{m}}\eta_{\mathrm{em}}(1 - \eta_{\mathrm{i}}) \tag{2.21}$$

in which:

$$\eta_{\mathrm{r}} = 1.666 - 0.004(t_{\mathrm{s}} - \Delta t_{\mathrm{o}}) - 0.00625(t_{\mathrm{u}} + \Delta t_{\mathrm{c}}) \tag{2.22}$$

$$\eta_{\mathrm{i}} = \left(0.425 + \frac{0.493Q_{\mathrm{HP}}}{1.16Q_{\mathrm{HP}} + 0.06}\right)\left(3.23 - 1.835\frac{t_{\mathrm{u}} + \Delta t_{\mathrm{c}}}{t_{\mathrm{s}} - \Delta t_{\mathrm{o}}}\right) \tag{2.23}$$

$$\eta_{\mathrm{m}} = 0.85 + \frac{0.158Q_{\mathrm{HP}}}{1.16Q_{\mathrm{HP}} + 0.1513\frac{t_{\mathrm{u}} + \Delta t_{\mathrm{c}}}{(t_{\mathrm{u}} + \Delta t_{\mathrm{c}}) - (t_{\mathrm{s}} - \Delta t_{\mathrm{o}})}} \tag{2.24}$$

$$\eta_{\mathrm{em}} = 0.85 + \frac{0.139Q_{\mathrm{HP}}}{1.335Q_{\mathrm{HP}} + 0.0904\frac{t_{\mathrm{u}} + \Delta t_{\mathrm{c}}}{(t_{\mathrm{u}} + \Delta t_{\mathrm{c}}) - (t_{\mathrm{s}} - \Delta t_{\mathrm{o}})}} \tag{2.25}$$

where t_{u} and t_{s} are the absolute temperatures of the hot and cold source, respectively; Δt_{c} is the temperature difference between the condensation temperature and the hot source temperature; Δt_0 is the temperature difference between the cold source temperature and the evaporation temperature; η_{r} is the efficiency of the real thermodynamic cycle compared to a reference Carnot cycle; η_{i} and η_{m} are the internal and mechanical efficiency of the compressor, respectively; η_{em} is the electromotor efficiency; and Q_{HP} is the thermal power of the HP.

To properly compare the performance of various HP types, the action energy must be uniform. In this sense, we calculate the useful heat delivered annually $Q_{\mathrm{u,year}}$ at annual equivalent fuel consumption $B_{\mathrm{fe,year}}$, necessary for driving power production, achieving the degree of fuel use φ_{year}, in kW/kg [6]:

$$\varphi_{\mathrm{year}} = \frac{Q_{\mathrm{u,year}}}{B_{\mathrm{fe,year}}} \tag{2.26}$$

The fuel economy depends on the HP type, according to Table 2.1.

Reducing GHG emissions, a key to limiting global warming, is associated with the replacement of traditional solutions for heating and cooling with HPs, particularly GSHPs. However, items related to electricity production, mainly used to drive them, must be taken into account.

Table 2.1 Energy Analysis of Heat Generation

No.	System Type	Fuel Use Degree	Primary Energy	Fuel Economy
		φ_{year} (kW/kg)	E_p (%)	ΔC (%)
1	Gas boiler	0.800	125.00	0
2	Heat pump with electro-compressor	1.083	92.34	− 32.66
3	Heat pump with electro-compressor and thermal boiler	0.969	103.20	− 21.80
4	Heat pump with thermal motor compressor	1.416	70.62	− 54.38
5	Absorption heat pump	1.219	82.03	− 42.97
6	Ejection heat pump	0.970	103.09	− 21.91

Table 2.2 Outdoor Air Temperatures for Different Partial Cooling Loads

Partial cooling load (%)	100	75	50	25
Outdoor air temperature (°C)	35	30	25	20

Currently, it is not recommended to replace a heating gas boiler with an electrically operated HP if the electricity is produced using coal or is based on old technologies because the resulting CO_2 emissions may increase by 1−2 t/year.

2.4.1.2 Partial Load Performances

In most of the annual cooling and heating periods, an HP is operated under the indoor thermal load lower than its rated capacity; this situation is called partial load operation.

The seasonal energy efficiency ratio (SEER) value in cooling and the seasonal coefficient of performance (SCOP) or heating seasonal performance factor (HSPF) values in heating reflect a HP's true energy consumption. SEER and SCOP measure the annual energy consumption and efficiency in typical day-to-day use. In the longer term, they take into account temperature fluctuations and standby periods to give a clear and reliable indication of the typical energy efficiency over an entire heating or cooling season. Seasonal efficiency indices indicates how an HP operates efficient over an entire cooling or heating season.

The term SEER is used in Europe to define the average annual cooling efficiency of an HP or A/C system. The term SEER is similar to the term EER but is related to a typical season rather than to a single-rated condition. The unit of SEER is Btu/(Wh). The SEER is a weighted average of EERs over a range of rated outdoor air conditions which can be calculated as:

$$\mathrm{SEER} = \frac{1 \cdot \mathrm{EER}_{100\%} + 42 \cdot \mathrm{EER}_{75\%} + 45 \cdot \mathrm{EER}_{50\%} + 12 \cdot \mathrm{EER}_{25\%}}{100} \tag{2.27}$$

where $\mathrm{EER}_{100\%}$, $\mathrm{EER}_{75\%}$, $\mathrm{EER}_{50\%}$ and $\mathrm{EER}_{25\%}$ are the energy efficiency ratios of HP at different partial cooling loads computed for the outdoor air temperatures mentioned in Table 2.2 and for the evaporator outlet water temperature of 6 or 7 °C and the temperature difference between supply and return cooled water of 5 °C.

ASHRAE Standard 90.1 uses the term integrated part-load value to report seasonal cooling efficiencies for both seasonal COPs and seasonal EERs (Btu/(Wh)), depending on the equipment capacity category.

The term SCOP is similar to the term SEER, except that it is used to signify the seasonal heating efficiency of HPs. The SCOP is a weighted average of COPs over a range of outside air conditions following a specific test method, according to European Standard EN 14825. The SCOP forms the basis for European minimum requirements and energy labelling for HPs.

Basically, the SCOP calculation method consists in dividing the heating season into a number of hours with different temperatures (called *bins*), which together are to reflect the variations in temperature over a heating season. Furthermore, a heating demand curve is determined for the temperatures, providing the heating demand that the HP is to meet for each set of temperatures. A COP value for each of the bins is found, and together these form the basis for calculating the average COP, i.e. the SCOP.

For the calculation of SCOP according to EN 14825, the HP must be tested at a series of points (temperatures), corresponding to the temperatures in EN 14511. At each of these temperatures, the HP can be tested for a more clearly defined partial load.

The test temperature on the cold side and the heating partial load of the HP for average climate zone are given in Table 2.3.

The SCOP method can be used to calculate both fixed- and variable-speed compressors. Similarly, full load data can be used to find the partial load efficiency values, if no partial load test results are available. The same temperature is applied when carrying out partial load as well as full load tests. If a COP has been found by use of a test for a capacity within a maximum of 10% above the calculated demand, according to EN 14825 this COP value may be used for this point. If a COP has been found at a lower capacity, this same COP value can also be used.

For fixed-speed compressors, COP and capacity are given at full load operation. Partial load is then adjusted using a degradation factor. This factor describes the reduction in energy efficiency at on/off operation. The factor can either be measured in a test or by using default values defined in EN 14825.

For air heating HPs (air-to-water and brine-to-water) with fixed-speed compressor, the EN 14825 assumes that the significant aspect for the degradation factor is the HP's remaining energy consumption when the compressor is turned off. Therefore, the energy consumption of the HP is measured after the compressor has been turned off for a minimum of 10 min. The degradation factor is subsequently calculated on the basis of this consumption figure as well as the full capacity of the HP at a similar test point. If this test has not been performed, a default value of 0.9 should be applied.

Table 2.3 Test Temperatures and Partial Load of Heat Pumps

Point	Partial Heating Load (%)	Temperatures (°C)
A	88	−7/20
B	54	2/20
C	35	7/20
D	15	12/20

Partial load COP for fixed capacity air-to-water and brine-to-water HPs at the test points is determined as follows:

$$COP_{\text{part load}} = COP_{DC}\frac{CR}{C_d CR + (1 - C_d)} \tag{2.28}$$

where COP_{DC} corresponds to COP at full load (declared capacity); C_d is the degradation factor; CR is the capacity ratio, that is the heating demand in relation to the HP's capacity at this temperature.

For HPs with capacity control, COP can be measured by testing at the stated capacity within an interval of 10%. If the HP has step control, partial load COP can be found by interpolation between COPs for the capacities above and below the desired capacity. If the desired capacity is below the minimum capacity of the HP, the calculation method for fixed-speed compressors should be used.

The calculated or measured partial load efficiency values are used when calculating the SCOP value. The calculation method includes finding a COP value for each temperature bin within the temperature interval of the climate zone. This COP gives the electricity consumption needed to meet the heating demand at each temperature. Interpolated values are used for temperatures between the determined test points. For temperatures outside the test points, extrapolation is performed using the two nearest test points. Electricity consumption is accumulated for all bins, including any necessary electric backup heating. Add to this, electricity consumption for standby and similar modes, in which the HP consumes a limited amount of electricity but not supplies the heating system. These types of consumption are determined on the basis of a specific number of hours in each mode.

By accumulating heating demand (E_t) and electricity consumption (E_{el}) for each temperature, $SCOP_{on}$ can be calculated as the accumulated heating demand divided by the accumulated electricity consumption (Eqn 2.12).

To calculate reference SCOP, the HP's electricity consumption when deactivated has to be included. This includes thermostat off mode, standby mode, crankcase heater mode and off mode. The electricity consumption for these modes is included in the calculation of the SCOP value as follows:

$$SCOP = \frac{E_t}{\frac{E_t}{SCOP_{on}} + H_{TO}E_{el,PO} + H_{SB}E_{el,SB} + H_{CK}E_{el,CK} + H_{OFF}E_{el,OFF}} \tag{2.29}$$

where H signifies the number of hours in a year the HP is in the stated operating mode, and E_{el} is the energy consumption of the HP in this mode, which is an input parameter in the calculation. The number of hours in each operating mode is determined on the basis of whether or not the HP can be used only for heating or for both heating and cooling.

2.4.1.3 Profitability and Capabilities of HP

The factors that can affect the life-cycle efficiency of an HP are (i) the local method of electricity generation; (ii) the local climate; (iii) the type of HP (ground or air source); (iv) the refrigerant used; (v) the size of the HP; (vi) the thermostat controls; and (vii) the quality of work during installation.

Considering that the HP has over-unit efficiency, evaluation of the consumed primary energy uses a synthetic indicator [4]:

$$\eta_s = \eta_g \mathrm{COP_{hp}} \tag{2.30}$$

in which:

$$\eta_g = \eta_p \eta_t \eta_{em} \tag{2.31}$$

where η_g is the global efficiency and η_p, η_t and η_{em} are the electricity production, the transportation and the electromotor efficiency, respectively.

To justify the use of an HP, the synthetic indicator has to satisfy the condition $\eta_s > 1$. Additionally, the use of an HP can only be considered if the $\mathrm{COP_{hp}} > 2.78$.

The COP of an HP is restricted by the second law of thermodynamics:

- in heating mode:

$$\mathrm{COP} \le \frac{t_u}{t_u - t_s} = \varepsilon_C \tag{2.32}$$

- in the cooling mode:

$$\mathrm{COP} \le \frac{t_s}{t_u - t_s} \tag{2.33}$$

where t_u and t_s are the absolute temperatures of the hot environment (condensation temperature) and the cold source (evaporation temperature), respectively, in K.

The maximum value ε_C of the efficiency can be obtained in the reverse Carnot cycle.

2.4.2 Economic Indicators

In the economic analysis of a system, different methods could be used to evaluate the systems. Some of them are: the present value (PV) method, the net present cost (NPC), the future value (FV) method, the total annual cost (TAC) method, the total updated cost (TUC) method, the annual life cycle cost (ALCC) and other methods [7,8].

- Present value analysis or discounting is often used to address the time value of money in project planning. With an appropriate discount rate, PV analysis translates a series of annual costs over a design lifetime for a project, of say 50 or 100 years, from the future to the present, enabling effects occurring at different times to be compared [9]. The PV of a future payment can be calculated using the equation:

$$\mathrm{PV} = \frac{C}{(1+i)^{\tau}} \tag{2.34}$$

where C is the payment/cost on a given future date; τ is the number of periods to that future date; i is the discount (interest) rate. Therefore, PV is the present value of a future payment that occurs at the end of the τ-th period.

Similarly, the PV of a stream of costs with a specified number of fixed periodic payments can be expressed as:

$$\mathrm{PV} = C \sum_{n=1}^{\tau} \left[\frac{1}{(1+i)^n} \right] \tag{2.35}$$

where C is the periodic payment that occurs at the end of each period; n is the number of periods (years).

The following equality can be demonstrated rather easily:

$$\sum_{n=1}^{\tau} \frac{1}{(1+i)^n} = \frac{(1+i)^\tau - 1}{i(1+i)^\tau} \tag{2.36}$$

and is defined update rate

$$u_\mathrm{r} = \frac{(1+i)^\tau - 1}{i(1+i)^\tau} = \frac{1}{\mathrm{CRF}} \tag{2.37}$$

where CRF is the capital recovery factor.

Taking into account Eqns (2.36) and (2.37), Eqn (2.35) yields the following equation:

$$\mathrm{PV} = u_\mathrm{r} C = \frac{C}{\mathrm{CRF}} \tag{2.38}$$

The discount rate i has a significant impact on the present value of a future cost. Generally, the higher the discount rate, the more reduction that occurs when a future value is converted into a PV.

The interest rate i can be calculated using the following equation [7]:

$$i = \left[\left(1 + \frac{P}{I_0}\right)^{1/q} - 1\right]^q - 1 \tag{2.39}$$

where $q = \log[1 + (1/N)]/\log 2$; N is the number of payment; P is the payment amount; I_0 is the initial investment cost.

- Net present cost of the project over its entire lifespan of operation includes expenses such as components, component replacements, operation and maintenance costs, and initial investment costs. The NPC can be computed using equation:

$$\mathrm{NPC} = \frac{\mathrm{TAC}}{\mathrm{CRF}} \tag{2.40}$$

where TAC is the total annual cost (sum of all annual costs of each system component).

- Another economic indicator is total updated cost:

$$\mathrm{TUC} = I_0 + \sum_{n=1}^{\tau} \frac{C}{(1+i)^n} \tag{2.41}$$

where I_0 is the initial investment cost in the operation, beginning the date of the systems inception; C is annual operation and maintenance cost of the system; i is the discount (inflation) rate; τ is the number of years for which is made update (20 years).

Taking into account Eqns (2.35) and (2.36), Eqn (2.41) yields the following equation:

$$\mathrm{TUC} = I_0 + u_\mathrm{r} C \tag{2.42}$$

- Usually the HP achieves a fuel economy ΔC (operating costs) comparatively of the classical system with thermal station (TS), which is dependent on the type of HP. On the other hand, the HP involves an additional investment I_{HP} from the classical system I_{TS}, which produces the same amount of heat.

Thus, it can be determined the recovery time RT, in years, to increase investment, $\Delta I = I_{HP} - I_{TS}$, taking into account the operation saving achieved through low fuel consumption $\Delta C = C_{TS} - C_{HP}$:

$$\mathrm{RT} = \frac{\Delta I}{\Delta C} \leq \mathrm{RT}_n \tag{2.43}$$

where RT_n is normal recovery time.

It is estimated that for RT_n a time frame within 8–10 years is acceptable, but this limit varies depending on the country's energy policy and environmental requirements.

- The TAC method can be used to compare the cost effectiveness of the HP over other heating systems. The ALCC is given by:

$$\mathrm{ALCC} = \mathrm{CRF} \cdot \mathrm{LCC} \tag{2.44}$$

where LCC is the life cycle cost which represents the cumulative cost of purchasing and running the system over its useful life and considers the influence of cost escalation on the annual operation and maintenance costs of the system. The LCC is given as:

$$\mathrm{LCC} = I_0 + \sum_{j=1}^{M} \frac{(1+e)^z}{(1+i)^z} \left(\frac{Q_{\mathrm{inc}} p_{\mathrm{f}}}{\mathrm{SPF}} + \mathrm{MC} \right) \tag{2.45}$$

where e is the annual fuel price escalation rate (%); i is the discount rate (%); z is the year after purchase of heating system (year); M is the total number of temperature bins (bin); Q_{inc} is the heating demand (kW); p_{f} is the base year fuel price (€/year); SPF is the seasonal performance factor (Btu/(Wh)); MC is the annual maintenance cost (€/year).

Considering both the interest and fuel cost escalation rates, the payback period of a GSHP compared with conventional heating sources would be the year that the PV of additional investment over that of the conventional source would equal the PV of the saving ΔC based on the first year of operation:

$$\mathrm{PV} = \Delta C \frac{\left(\frac{1+e}{1+i}\right)^b - 1}{1 - \frac{1+e}{1+i}} \tag{2.46}$$

where b is the year to payback [8].

The assumptions made in the economic comparisons of an HP with the other conventional heating systems are given in Table 2.4 [10]. Using a lifetime of 20 years for all the heating systems, an interest rate of 8% and an annual fuel price escalation rate of 4%, Esen et al. [10] calculated the ALCC of each heating system for the prices of the fuels in Turkey (based on 2006 conditions). For the assumed fuel prices, natural gas has the lowest annual life cycle cost, closely followed by the GSHP system. The GSHP is more cost effective than the heating systems using electric resistance, fuel oil, liquid petrol gas, coal, and Diesel oil.

From Eqn (2.45), the payback period of the GSHP would be 8.38 years against the electric resistance, 23.17 years against the fuel oil, 12.43 years against the liquid petrol gas, 35.68 years against the natural gas, 20.75 years against the coal and 10.31 years against the Diesel oil. If the seasonal performance of the analysed GSHP system is

Table 2.4 Assumptions for Each Fuel

Fuels and GSHP	Average Heating Value	Average Efficiency	Price of Fuels	Consumption of Fuel Daily for Heating Demand
Natural gas	8250 kcal/m^3	91%	0.154 €/m^3	3.18 m^3/day
Coal	6000 kcal/kg	69%	0.146 €/kg	4.38 kg/day
Fuel oil	9700 kcal/kg	81%	0.249 €/kg	2.71 kg/day
Liquid petrol gas	11000 kcal/m^3	91%	0.592 €/m^3	2.39 m^3/day
Oil	10200 kcal/kg	85%	0.618 €/kg	2.57 kg/day
Electric resistance	860 kcal/kWh	99%	0.074 €/kWh	30.6 kWh/day
GSHP	860 kcal/kWh	2.74 (COP_{sys})	0.074 €/kWh	30.6 kWh/day

improved or its capital cost is reduced, the GSHP would be much more economic than the other heating systems.

2.4.3 Calculation of GHG Emissions

Due to the diversity in each country with respect to heating practices, direct geothermal energy use by GSHPs, and primary energy sources for electricity, country-specific calculations are provided.

The annual heating energy provided by GSHPs is defined as E_t. The annual primary energy consumption from HP electricity use is then:

$$E_{el} = \frac{E_t}{SPF} \tag{2.47}$$

Because HP electricity consumption is considered the most important source for GHG emissions [11], other potential contributors (e.g. HP life cycle, HP refrigerant and borehole construction) are neglected. Applying an emission factor g_p, in kg CO_2/kWh, the annual GHG emissions C_{GSHP}, in kg CO_2, from GSHP operation can be obtained:

$$C_{GSHP} = g_p E_{el} \tag{2.48}$$

The emission factor typically varies among different countries and characterises the GHG intensity of electricity production. Note that although carbon dioxide (CO_2) represents the most important GHG, there are several other compounds that contribute similarly to climate change. Their combined impact is commonly normalised to the specific effect of CO_2, and all emissions are expressed in CO_2 equivalents. For the sake of readability, however, the emissions are expressed only in kilograms of CO_2.

Thus, the CO_2 emissions C_{CO_2} of the GSHP during its operation can be evaluated with the following equation:

$$C_{CO_2} = g_{el} E_{el} \tag{2.49}$$

where g_{el} is the specific CO_2 emission factor for electricity. The average European CO_2 emission factor for electricity production is 0.486 kg CO_2/kWh and for Romania is 0.547 kg CO_2/kWh [12].

Theoretical emissions C_{sub}, in kg CO_2, from the substituted energy by GSHP are determined by E_t and the emission factor g_f representative for the substituted heat mix:

$$C_{sub} = g_f E_t \quad (2.50)$$

The substituted heat is a mix from different energy carriers, i. The emission factor thus depends on the portion $e_{sub,i}$ of each energy carrier in the substituted heat mix:

$$g_{sub} = \sum_i g_{f,i} e_{sub,i} \quad (2.51)$$

The portions $e_{sub,i}$ ($\Sigma_i e_{sub,i} = 1$) are also termed substitution factors [11].

The annually saved emissions ΔC_{GHG} are obtained by:

$$\Delta C_{GHG} = C_{sub} - C_{GSHP} \quad (2.52)$$

According to Eqn (2.52), $C_{GHG} = 0$ indicates no savings and negative values denote increased GHG emissions from GSHPs in comparison to those from conventional heating.

2.5 ENVIRONMENTAL IMPACT

The US Environmental Protection Agency (EPA) has called ground-source HPs the most energy-efficient, environmentally clean and cost-effective space conditioning systems available. HPs offer significant emission reductions potential, particularly where they are used for both heating and cooling and where the electricity is produced from renewable resources.

Due to the generally high COP of an HP and the utilisation of solar and geothermal energy stored in the subsurface, GSHP systems are capable of lower CO_2 emissions than other conventional heating methods, such as oil-fired heating. Thus, the use of GSHPs for heating and cooling of residential and commercial buildings can significantly reduce the emissions of global GHGs such as CO_2 and SO_2.

Ground-source HPs have unsurpassed thermal efficiencies and produce zero emissions locally, but their electricity supply includes components with high GHG emissions, unless the owner has opted for a 100% renewable energy supply (wind, hydro, photovoltaic or solar thermal). The environmental impact of HPs therefore depends on the characteristics of the electricity supply and the available alternatives.

The GHG emissions savings ΔC_{GHG}, in t CO_2/year, from an HP over a conventional furnace can be obtained from Eqn (2.52):

$$\Delta C_{GHG} = E_t \left(\frac{g_f}{1000 \varepsilon_f} - \frac{g_{el}}{3600 \mathrm{SPF}} \right) \quad (2.53)$$

where E_t is the seasonal heating energy, in GJ/year; g_f is the emission factor for fuel, in kg CO_2/GJ; ε_f is the furnace efficiency; g_{el} is the emission factor for electricity, in t CO_2/GWh, depending on the region; and SPF is the seasonal performance factor of the HP.

Ground-source HPs always produce fewer GHGs than A/Cs, oil boilers, and electric heaters, but natural gas boilers may be competitive depending on the GHG intensity of the local electricity supply.

Several studies have reported reduced and avoided emissions compared to conventional, fossil-fuel-based heating technologies. A European study showed that electrically driven HPs avoid additional CO_2 emissions by 45% compared with an oil boiler and

33% compared with a gas-fired boiler. A regional analysis based on a subsidy program of GSHP was performed in southwestern Germany using a geographic information system, showing the regional achieved CO_2 savings. The emitted CO_2 per kWh of heating demand for the studied scenario resulted in 0.149 kg CO_2/kWh for GSHP using the German electricity mix and 0.065 kg CO_2/kWh using the regional electricity mix, which results in CO_2 savings of 35% or 72%, respectively [13]. The calculated savings vary substantially and, in particular, depend on the case-specific electricity mix for running the HP. Saner et al. [14] showed that when contrasting the conditions in different European countries, emission savings of up to 88% are possible, with a median value of about 35% compared to oil-fired boilers. In their study on the Sanmartin village in Romania, Blaga et al. [15] discuss the environmental and economic benefits of replacing the mainly wood-based single-house heating systems with a shallow geothermal district heating system. They not only identify significant advantages with respect to CO_2 reductions from operating open systems (GWHPs) but also emphasise the decrease of other combustion and gas ash emissions, as well as reductions in the residual heat loss from releasing combustion gases to the air. By operating heat exchangers based on plates, the overall environmental impacts are predicted to be close to zero.

Since one main determinant of the environmental performance of GSHPs is the SFP, and variability of representative values exists depending on mainly climate, technology and innovative progress, a range between low (SPF = 3) and high efficiency (SPF = 4) was considered by Bayer et al. [11]. Figure 2.5 shows for all selected countries (except of Estonia, Spain and Hungary), that in fact this SPF range has a large influence on potential GHG savings. This is reflected by the width of the grey bars that visualise the difference in the figure. It is also shown that in countries with very carbon-intense electricity production, a SPF of 3 often is not sufficient.

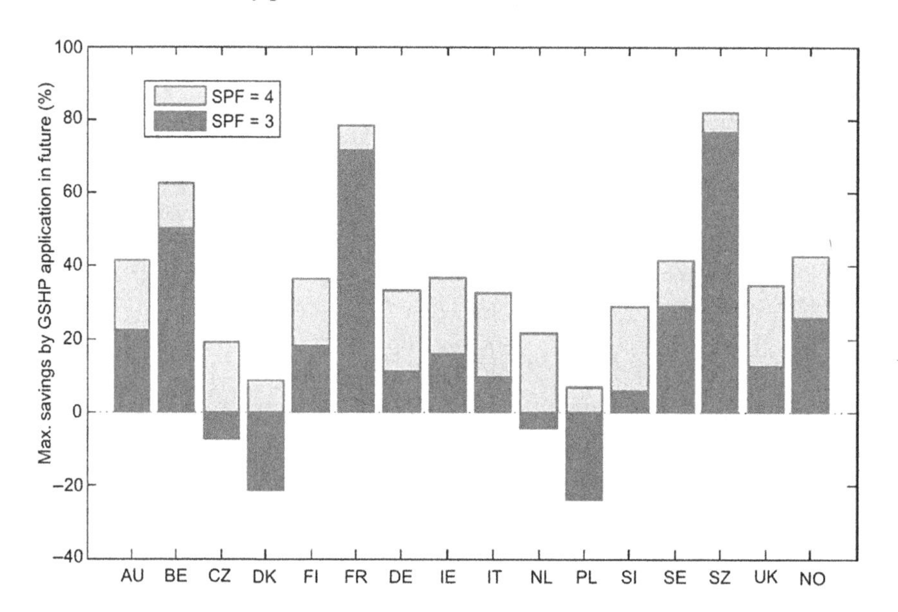

Figure 2.5 Maximum GHG savings by GSHPs in European countries by a complete replacement of conventional heating in the residential sector.

For Czech Republic, Poland and the Netherlands, even increased GHG emissions would be calculated. This is also the case for Denmark, where the electricity emission rate is not extreme, but in this country, household heatings reveal to be dominated (e.g. by wood-fired boilers) and thus benefits can hardly be achieved from GSHP operation. For these four countries, savings are possible by increased SPF values. However, due to the primary energy consumption by the HPs even complete replacement of conventional heating technologies with GSHPs would only yield savings of about 20% or less of the current GHG emissions caused by heating. For most other countries, the potential savings lie between 20% (SPF = 3) and 40% (SPF = 4). Countries with electricity mixes dominated by nuclear and hydropower production show higher potential saving rates. Saving rates reach between 60% and 80% for Belgium, France and Switzerland. Sweden reveals to be well developed already, and thus, relative future savings are limited and range between 10% and 40%. Summing the maximum savings together, at most about 150 Mio t CO_2 emissions (SPF = 3.5) could be saved for the studied countries (based on 2008 conditions), which represents a mean value of 30% (19–38%) [11].

Another study performed by Saner et al. [14] showed the environmental benefits from GSHP systems by computing the CO_2 emission savings compared to conventional heating (and cooling) systems, such as oil burner or gas furnaces. A comparison is made without taking into account the technology-specific requirements of the heating system (e.g. suggesting the combination of radiant floor heating and GSHP). Heat production in cold seasons is combined with a specified rate of cooling power output during the warm season. A GSHP system is shown to supply a single central European family house with a peak heat demand of 10 kW. In addition, the cooling power output for passive cooling in summer is assumed to be 5 kW.

In the base case, the heat is extracted from the ground by means of two borehole heat exchangers with a depth of 85 m each. The boreholes are drilled by flush or sledge hammer techniques, consuming 1.5 or 2.5 l of diesel oil per metre, respectively. During operation, the GSHP system needs low voltage electricity to run the HP refrigeration cycle and circulate the heat carrier liquid. The average electricity mix of continental Europe with a carbon footprint of 0.599 kg CO_2/kWh is used.

The percentage values listed in Table 2.5 show CO_2 savings from selecting a GSHP system instead of a conventional heating system, without considering cooling services. Negative values denote cases where more CO_2 was emitted with a GSHP system than with a conventional heating system. The values in brackets are the CO_2 savings that can be achieved when including the effect of passive cooling provided by GSHP systems compared to a conventional heating system, which is expanded to include the provision of cooling via an air-conditioning system.

The CO_2 savings for Europe, which are between 31% and 88% compared to conventional heating systems, such as oil-fired boilers and gas furnaces, largely depend on the primary resource of the supplied electricity for the HP, the climatic conditions and the inclusion of passive cooling capabilities.

Open loop systems (i.e. those that draw ground-water as opposed to closed-loop systems using a borehole heat exchanger) need to be balanced by reinjecting the spent water. This prevents aquifer depletion and the contamination of soil or surface water with brine or other compounds from underground according to the Directive 2006/

Table 2.5 Savings of CO_2 Emissions for GSHP System Compared with Oil Boiler and Gas Furnace Heating

Country	Code	Oil Boiler Heating (and Air-Condition Cooling)	Gas Furnace Heating (and Air-Condition Cooling)
Austria	AT	56(58)	53(54)
Bosnia	BA	19(24)	13(18)
Belgium	BE	62(63)	48(49)
Bulgaria	BG	30(33)	25(29)
Switzerland	CH	81(81)	76(76)
Czech Republic	CZ	20(24)	12(17)
Germany	DE	35(38)	21(25)
Denmark	DK	43(45)	22(26)
Spain	ES	45(47)	31(34)
Finland	FI	64(65)	62(63)
France	FR	82(83)	79(79)
United Kingdom	UK	38(41)	12(17)
Greece	GR	4(10)	−2(4)
Croatia	HR	45(47)	40(43)
Hungary	HU	33(36)	28(32)
Ireland	IE	23(27)	−6(2)
Italy	IT	41(44)	28(32)
Luxembourg	LU	41(44)	19(23)
Montenegro	ME	21(25)	15(20)
Macedonia	MK	−5(2)	−14(−6)
Netherlands	NL	35(38)	9(14)
Norway	NO	87(87)	83(83)
Poland	PL	1(8)	−21(−12)
Portugal	PT	37(40)	21(25)
Romania	RO	29(32)	20(24)
Serbia	RS	−1(5)	−15(−6)
Sweden	SE	83(83)	76(76)
Slovenia	SI	53(55)	50(52)
Slovakia	SK	52(53)	49(50)
Median value		38(41)	22(26)

118/EC of the European Parliament and of the Council of 12 December 2006 on the protection of groundwater against pollution and deterioration. The Directive 2000/60/EC also requires EU member states to implement the necessary measures to prevent deterioration of surface water states and to prevent or limit the introduction of pollutants into groundwater.

REFERENCES

[1] Cengel Y, Boles M. Thermodynamics: an engineering approach. New York, NY: McGraw Hill; 2014.

[2] Sarbu I, Sebarchievici C. General review of ground-source heat pump system for heating and cooling of buildings. Energy Build 2014;70(2):441–54.

[3] Heinonen EW, Tapscott RE, Wildin MW, Beall AN. NMERI 96/15/32580 Assessment of antifreeze solutions for ground-source heat pump systems. New Mexico Engineering Research Institute; 1996

[4] Sarbu I, Sebarchievici C. Heat pumps. Timisoara: Polytechnic Publishing House; 2010 [in Romanian]

[5] Radcenco V, Florescu AL, Duica T, Burchiu N, Dimitriu S, et al. Heat pumps systems. Bucharest: Technical Publishing House; 1985 [in Romanian]

[6] Sebarchievici C. Optimization of thermal systems from buildings to reduce energy consumption and CO_2 emissions using ground-coupled heat pump [Doctoral thesis]. Romania: Polytechnic University Timisoara; 2013.

[7] Thuesen GJ, Fabrycky WJ. Engineering economy. Englewood Cliffs, NJ: Prentice-Hall International Editions; 1989.

[8] Tassou SA, Maequand CJ, Wilson DR. Energy and economic comparisons of domestic heat pumps and conventional heating systems in the British Climate. Appl Energy 1986;34(2):127–38.

[9] Kaen FR. Corporate finance: concepts and policies. USA: Blackwell Business; 1995.

[10] Esen H, Inalli M, Esen M. Technoeconomic appraisal of a ground source heat pump system for a heating season in eastern Turkey. Energy Convers Manage 2006;47(9–10):1281–97.

[11] Bayer P, Saner D, Bolay S, Rybach I, Blum P. Greenhouse gas emission savings of ground source heat pump systems in Europe. Renewable Sustainable Energy Rev 2012;16:1256–67.

[12] IEE, Intelligent Energy Europe, <http://ec.europa.eu/energy/environment>; 2013.

[13] Blum P, Campilo G, Münch W, Kölbel T. CO_2 savings of ground source heat pump systems – a regional analysis. Renewable Energy 2010;35:122–7.

[14] Saner D, Juraske R, Kubert M, Blum P, Hellweg S, Bayer P. Is it only CO_2 that maters? A life cycle perspective on shallow geothermal systems. Renewable Sustainable Energy Rev 2010;14:1798–813.

[15] Blaga AC, Rosca M, Karytsas K. Heat from very low enthalpy geothermal source versus solid fuels in the Felix-Sanmartin area, Romania. In: 2010 world geothermal congress, Bali, Indonesia, April 25–30; 2010.

CHAPTER 3

Substitution Strategy of Non-Ecological Refrigerants

3.1 GENERALITIES

Environmental pollution represents a major risk for all life on our planet (e.g. humans, flora and fauna), because it consists not only of the local noxious effects of different pollutants, but also the imbalances produced on a large scale over the entire planet. Environmental protection represents the fundamental condition of the society's sustainable development and is a high priority of national interest realised via institutional framework in which the legal norms regulate the development of activities with environmental impact and exert control over such activities.

The purpose of environmental protection is to maintain the ecological balance, to maintain and improve the natural factors, to prevent and control pollution, to promote the development of natural values, to ensure better life and work conditions for the present and future generations, and it refers to all actions, means and measures undertaken for these purposes.

One of the minor components of the atmosphere, the ozone layer, has a special importance in maintaining the ecological balance. Ozone is distributed primarily between the stratosphere (85–90%) and troposphere. Any perturbation of the atmospheric ozone concentration (which varies between 0 and 10 ppm, depending on the regions) has direct and immediate effects upon life.

For most of the states the problems of forming and maintaining the earth's ozone layer, represents a major priority. In this context during the past 30 years, the European Union has adopted a large number of laws and regulations concerning environmental protection to correct the pollution effects, frequently by indirect directives, through imposition of the levels of allowable concentrations by asking for government collaboration, programs and projects for the regulation of industrial activities and productions. The Alliance for Responsible Atmospheric Policy is an industry coalition and leading voice for ozone protection and climate change policies, which maintains a brief summary of the regulations for some countries [1].

Refrigerants are the working fluids in heat pump (HP), air-conditioning (A/C), and refrigeration systems. They absorb heat from one area, such as an air-conditioned space, and reject it into another, such as one outdoors, usually through evaporation and condensation.

Working fluids that escape through leakages from cooling equipment during normal operation (filling or empting) or after accidents (damages) gather in significant quantities at high levels of the atmosphere (stratosphere). In the stratosphere, through catalytically decompounding, pollution from working fluid leakage depletes the ozone

Ground-Source Heat Pumps. DOI: http://dx.doi.org/10.1016/B978-0-12-804220-5.00003-5

layer that normally filters the ultraviolet radiation from the sun, which is a threat to living creatures and plants on earth. Stratospheric ozone depletion has been linked to the presence of chlorine and bromine in the stratosphere. In addition, refrigerants contribute to global warming (also called global climate change) because they are gases that exhibit the greenhouse effect when in the atmosphere.

Concerning the polluting action upon the environment, for atmospheric ozone, as presented through the Montreal Protocol [2] and the subsequent amendments, as well as for the greenhouse effect according to the Kyoto Protocol [3], refrigerants can be classified as follows:

- having strong destructive action on the ozone layer and with significant amplification of the greenhouse effect upon the earth (chlorofluorocarbons – CFCs)
- having reduced action on the ozone layer and with moderate amplification of the greenhouse effect (hydro-chlorofluorocarbons – HCFCs)
- being harmless to the ozone layer, with less influence upon the greenhouse effect (hydro-fluorocarbons – HFCs)
- being harmless to the ozone layer, with very less or even no influence upon greenhouse effect (carbon dioxide – CO_2 (R744), natural hydrocarbons (HCs) and ammonia – NH_3 (R717) respectively).

Vapour compression-based systems are generally employed in HP units operating with halogenated refrigerants. The international protocols (Montreal and Kyoto) restrict the use of the halogenated refrigerants in the vapour compression-based systems. As per the 1987 Montreal Protocol, the use of CFCs was completely stopped in most of the nations. However, HCFCs refrigerants can be used until 2040 in developing nations and developed nations should phase usage out by 2030 [4]. To meet the global demand in the HP and A/C sector, it is necessary to look for long-term alternatives to satisfy the objectives of international protocols. From the environmental, ecological and health point of view, it is urgent to find some better substitutes for HFC refrigerants [5]. HC and HFC refrigerant mixtures with low environment impacts are considered potential alternatives to phase out the existing halogenated refrigerants.

This chapter presents a study on the recent development of possible substitutes for non-ecological refrigerants employed in heating, ventilating, air-conditioning and refrigerating (HVAC&R) equipment based on thermodynamic, physical and environmental properties and total equivalent warming impact (TEWI) analysis. This study contains a good amount of information regarding the environmental pollution produced by the working fluids of the HP, A/C, and commercial refrigeration applications and the ecological refrigerant trend. Overall, it is useful for readers who are interested in the current status of alternative refrigerant development related to vapour compression-based systems. The study describes the selection of refrigerants adapted to each utilisation based on the thermodynamic, physical and environmental properties, the technological behaviour and the usage constraints as the principal aspects of environmental protection. Also, this chapter explores the studies reported with new refrigerants in HPs, domestic and commercial refrigerators, chillers and in automobile A/Cs.

3.2 ENVIRONMENTAL IMPACT OF REFRIGERANTS

The design of the refrigeration equipment depends strongly on the properties of the selected refrigerant. Refrigerant selection involves compromises between conflicting desirable thermo-physical properties. A refrigerant must satisfy many requirements, some of which do not directly relate to its ability to transfer heat. Chemical stability under conditions of use is an essential characteristic. Safety codes may require a non-flammable refrigerant of low toxicity for some applications. The environmental consequences of refrigerant leaks must also be considered. Cost, availability, efficiency and compatibility with compressor lubricants and equipment materials are other concerns.

Safety properties of refrigerants considering flammability and toxicity are defined by ASHRAE Standard 34 [6]. Toxicity classification of refrigerants is assigned to classes A or B (Table 3.1). Class A signifies refrigerants for which toxicity has not been identified at concentrations less than or equal to 400 ppm by volume, and class B signifies refrigerants with evidence of toxicity at concentrations below 400 ppm by volume. By flammability refrigerants are divided in three classes. Class 1 indicates refrigerants that do not show flame propagation when tested in air (at 101 kPa and 21 °C). Class 2 signifies refrigerants having a lower flammability limit (LFL) of more than 0.10 kg/m^3 and a heat of combustion less than 19,000 kJ/kg. Class 3 indicates refrigerants that are highly flammable, as defined by a LFL of less than or equal to 0.10 kg/m^3 or a heat of combustion greater than or equal to 19,000 kJ/kg.

New flammability class 2L has been added since 2010 and denotes refrigerants with burning velocities less than 10 cm/s.

Minimising all refrigerant releases from systems is important not only because of environmental impacts, but also because charge losses lead to insufficient system charge levels, which in turn results in suboptimal operation and lowered efficiency.

The average global temperature is determined by the balance of energy from the sun heating the earth and its atmosphere and of the energy radiated from the earth and the atmosphere into space. Greenhouse gases (GHGs), such as water vapour, as well as small particles trap heat at and near the surface, maintaining the average temperature of the Earth's surface at a temperature approximately 34 K warmer than if these gases and particles were not present (this phenomenon is referred to as the greenhouse effect).

Global warming is a concern because of an increase in the greenhouse effect from increasing concentrations of GHGs, which are attributed to human activities. Thus, the negative environmental impact of the working fluids, especially the effect of halogenated refrigerants on the environment, can be synthesised by two effects [7]:

- depletion of the ozone layer
- contribution to global warming at the planetary level via the greenhouse effect.

The measure of a material's ability to deplete stratospheric ozone is its *ozone depletion potential* (ODP), a relative value to that of R11, which has an ODP of 1.0.

Table 3.1 Safety Classification of Refrigerants

Flammability	Safety Code	
	Lower Toxicity	Higher Toxicity
Higher flammability	A2	B2
Lower flammability	A2L	B2L
No flame propagation	A1	B1

The global warming potential (GWP) of a GHG is an index describing its relative ability to collect radiant energy compared with CO_2, which has a very long atmospheric lifetime. Therefore, refrigerants will be selected so that the ODP will be zero with a reduced GWP.

The most utilised halogenated refrigerants are the family of chemical compounds derived from the HCs (methane and ethane) by substitution of chlorine (Cl) and fluorine (F) atoms for hydrogen (H), whose toxicity and flammability scale according to the number of Cl and H atoms. The presence of halogenated atoms is responsible for ODP and GWP.

Table 3.2 presents the principal characteristics of halogenated refrigerants (pure and mixtures), with the symbol for refrigerant, chemical name and formula, as well as their application domains [8].

During the last century, the halogenated refrigerants have dominated the vapour compression-based systems due to their good thermodynamic and thermo-physical properties. Thermodynamic properties of pure refrigerants are listed in Table 3.3 [9]. The halogenated refrigerants have poor environmental properties with respect to ODP and GWP.

The second generation of refrigerants, CFCs, replaced classic refrigerants in the early twentieth century. CFCs (R12, R11 and R13) have been used as refrigerants since the 1930s because of their superior safety and performance characteristics. However, their production for use in developed countries has been eliminated because they deplete the ozone layer. The CFCs and HCFCs represented by R22 and mixture R502 dominated the second generation of refrigerants.

HCFCs also deplete the ozone layer, but to a much lesser extent than CFCs. HCFCs production for use as refrigerants is scheduled for elimination by 2030 for developed countries and by 2040 for developing countries [4].

The traditional refrigerants (CFCs) were banned by the Montreal Protocol because of their contribution to the stratospheric ozone layer's disruption. The Kyoto Protocol listed HCFCs as having large GWPs.

With the phasing out of the use of CFCs, chemical substances, such as the HCFCs and the HFCs, were proposed and have been used as temporary alternatives.

The HFCs do not deplete the ozone layer and have many of the desirable properties of CFCs and HCFCs. They are being widely used as substitute refrigerants for CFCs and HCFCs. The HFC refrigerants have significant benefits regarding safety, stability and low toxicity, and are appropriate for large-scale applications.

Also, the HC and HFC refrigerant mixtures with low environment impacts are considered as potential alternatives to phase out the existing halogenated refrigerants. HC-based mixtures are environment-friendly, and can be used as alternatives without modifications in the existing systems. However, HC refrigerant mixtures are highly flammable, which limits the usage in large capacity systems [10]. HFC mixtures are ozone-friendly, but have significant GWP. HFC mixtures are not miscible with mineral oil, which requires synthetic lubricants (such as polyolester). Earlier investigations reported that HFC/HC mixtures are miscible with mineral oil. It is possible to mix HC refrigerants with HFC to replace the existing halogenated refrigerants [11].

A second influence of refrigerants upon the environment, as previously mentioned, led to a new classification of refrigerants according to their contributions to global warming. Comparison of these specific contributions to the greenhouse effect is

Table 3.2 Application Domains of Halogenated Refrigerants

Group	Refrigerant	Chemical Formula/Chemical Name	Evaporation Temperature, t_0 (°C)		Applications
0	1	2	3	4	5
CFC	R11	CCl_3F Threechlorofluoromethane	0	+60	Air-conditioning, heat pumps
	R12	CCl_2F_2 Dichlorodifluoromethane	−40 +10	+10 +40	Domestic and commercial refrigeration Air-conditioning, heat pumps
	$R12B_1$	$CClBrF_2$ Bromochlorodifluoromethane	0	+50	Air-conditioning, heat pumps
	R13	$CClF_3$ Chlorotrifluoromethane	−100	−60	Cascade refrigeration systems
	$R13B_1$	$CBrF_3$ Bromotrifluoromethane	−80	−40	Mono-, two stage- and in cascade refrigeration systems, for industry
	R113	$C_2Cl_3F_3$ Trichlorotrifluoroethane	0 +15	+15 +50	Air-conditioning Heat pumps
	R114	$C_2Cl_2F_4$ Dichlorotetrafluoroethane	−20 +10	+10 +80	Air-conditioning Heat pumps
HCFC	R21	$CHCl_2F$ Dichlorofluoromethane	−20	+20	Air-conditioning, heat pumps
	R22	$CHClF_2$ Chlorodifluoromethane	−50	+10	Industrial-, food-, commercial refrigeration, air-conditioning
	R142b	$C_2H_3ClF_2$ Chlorodifluoroethane	−20 +10	+10 +60	Air-conditioning Heat pumps
HFC	R23	CHF_3 Threefluoromethane	−100	−60	Cascade refrigeration systems for industry and laboratory
	R32	CH_2F_2 Difluoromethane	−60	−10	Industrial and commercial refrigeration
	R125	C_2HF_5 Pentafluoroethane	−50	+10	Industrial and commercial refrigeration, air-conditioning
	R134a	$C_2H_2F_4$ Tetrafluoroethane	−30	+20	Domestic-, commercial-, industrial refrigeration, air-conditioning
	R152a	$C_2H_4F_2$ Difluoroethane	−30	+10	Industrial and commercial refrigeration, air-conditioning
Mixtures	R500	(R12/R152a)	−40	+10	Household and industrial refrigeration, heat pumps
	R502	(R22/R115)	−60	−20	Industrial and commercial refrigeration
	R507	(R125/R134a)	−50	−10	Industrial and commercial refrigeration
	R410A	(R32/R125)	−50	0	Industrial and commercial refrigeration
	R407C	(R32/R125/R134a)	−40	0	Industrial and commercial refrigeration
	R404A	(R125/R143a/R134a)	−40	0	Industrial and commercial refrigeration

Table 3.3 Thermodynamic Properties of Pure Refrigerants

Refrigerant	Molecular Mass, M (g/mol)	Critical Temperature, t_{cr} (°C)	Critical Pressure, p_{cr} (MPa)	Boiling Point, t_{0n} (°C)
R11	137.37	198.0	4.41	23.7
R12	120.90	112.0	4.14	−29.8
R22	86.47	96.2	4.99	−41.4
R23	70.01	25.9	4.84	−82.1
R32	52.02	78.2	5.80	−51.7
R41	34.03	44.1	5.90	−78.1
R123	152.93	82.0	3.66	27.8
R124	136.48	122.3	3.62	−12.0
R125	120.02	66.2	3.63	−54.6
R134a	102.03	101.1	4.06	−26.1
R142b	100.49	137.2	4.12	−9.0
R143a	84.04	72.9	3.78	−47.2
R152a	66.05	113.3	4.52	−24.0
R161	48.06	102.2	4.70	−34.8
R170	30.07	90.0	4.87	−88.9
R218	188.02	71.9	2.68	−36.6
R290	44.10	96.7	4.25	−42.2
R600	58.12	152.0	3.80	−0.5
R600a	58.12	134.7	3.64	−11.7
R717	17.03	132.3	11.34	−33.3
R744	44.01	31.1	7.38	−78.4
R1270	42.08	92.4	4.67	−47.7

performed for R11 (the most noxious even from the point of view of ODP) as well as for CO_2. Halogenated refrigerants are categorised in the undesirable position 3 (14%) between the GHGs, perhaps due to their great absorption capacity for infrared radiation.

In the case of HP systems, although supplementary to the direct action to the greenhouse effect because of the refrigerants' leakage in atmosphere, it must be considered even the indirect action to global warming by the CO_2 quantity released during the production of the drive energy for the system is obviously greater than the associated direct action [12]. While the refrigerant quantity increases in the system, the effect of direct action rises.

The environmental impact of an HVAC&R system is due to the release of refrigerant and the emission of GHGs for associated energy use. TEWI is used as an indicator for environmental impact of the system for its entire lifetime. TEWI is the sum of the direct refrigerant emissions, expressed in terms of CO_2 equivalents, and the indirect emissions of CO_2 from the system's energy use over its service life.

The *life-cycle climate performance* (LCCP) of an HVAC&R system includes TEWI and adds the effects of direct and indirect emissions associated with manufacturing the refrigerant. The analysis of the TEWI index for refrigeration systems operating with different refrigerants (i.e. R22, R134a, R404A, R717, R744) indicated that the direct effect generated by CO_2 is negligible compared with the other refrigerants [13]. The indirect effect generated by CO_2 is significant because of the high condensation pressures that determine the large amount of energy consumption and, consequently, the maximum value of TEWI for CO_2.

Environmentally preferred refrigerants have:

- low or zero ODP
- relatively short atmospheric lifetimes
- low GWP
- ability to provide good system efficiency
- appropriate safety properties
- ability to yield a low TEWI or LCCP in system applications.

Table 3.4 lists the environmental properties of refrigerants [8]. Because HFCs do not contain chlorine or bromine, their ODP values are negligible and represented by 0 in this table. NH_3, HCFCs, most HFCs, and HFOs have shorter atmospheric lifetimes than CFCs because they are largely destroyed in the lower atmosphere by reactions with OH radicals. A shorter atmospheric lifetime generally results in lower ODP and GWP values.

The European Commission [14] has published its firm proposal for changes to EU F-Gas Regulation. These changes aim to substantially reduce the emissions of fluorinated (F) gases over the next 20 years. F-gases are GHGs, with a GWP several thousand times higher that of CO_2. The largest use of F-gases in Europe is for HFC refrigerants, and under the new F-gas regulation, their use will be severely restricted. The two most important proposals are:

- a phase down in EU consumption of HFCs via a series of cuts starting in 2016
- a ban on the use of high-GWP refrigerants from 2020.

3.3 INFLUENCE OF REFRIGERANTS ON PROCESS EFFICIENCY

The design and efficiency of the refrigeration equipment depends strongly on the selected refrigerant's properties. Consequently, operational and equipment costs significantly depend on the refrigerant choice. The single-stage vapour-compression system with a single component or azeotropic refrigerant has the thermodynamic cycle illustrated in Figure 2.2.

When zeotropic mixtures are used as refrigerants, gliding temperatures influence cycle efficiency as well as system design.

Temperature glide appears during evaporation and condensation at constant pressure. Use of counter flow heat exchangers can sometimes help to utilise that temperature glide efficiently, but problems can appear with leakage of refrigerants from such systems as the initial refrigerant composition, and thus properties can be disturbed.

Comparison of different refrigerants gives a good overview of achievable cycle performance for a basic referent cycle [15]. Table 3.5 compares refrigerants' reference

Table 3.4 Environmental Properties of Refrigerants

Group	Fluid	ODP	GWP (R11 = 1)	GWP (CO_2 = 1)	Atmospheric Lifetime (years)
CFC	R11	1	1	4000	50–60
	R12	1	2.1–3.05	10,600	102–130
	R113	0.8–1.07	1.3	4200	90–110
	R114	0.7–1.0	4.15	6900	130–220
	$R12B_1$	3–13	–	1300	11–25
	$R13B_1$	10–16	1.65	6900	65–110
HCFC	R21	0.05	0.1	–	<10
	R22	0.055	0.034	1900	11.8
	R123	0.02	0.02	120	1.4–2
	R142b	0.065	0.3–0.46	2000	19–22.4
HFC	R23	0	6	14,800	24.3
	R32	0	0.14	580	6–7.3
	R125	0	0.58–0.85	3200	32.6
	R134a	0	0.28	1600	14–15.6
	R143a	0	0.75–1.2	3900	55–64.2
	R152a	0	0.03–0.04	140	1.5–8
HFO	R1234yf	0	–	<4.4	0.029
NH_3	R717	0	–	0	<0.02
CO_2	R744	0	–	1	>50
Azeotropic mixtures	R500(R12/R152a)	0.63–0.75	2.2	6000	–
	R501(R12/R22)	0.53	1.7	4200	–
	R502(R22/R115)	0.3–0.34	4.01–5.1	5600	>100
	R507(R125/R143a)	0	0.68	3800	–
Near azeotropic mixtures	R404A(0.44R125/0.52R143a/0.04R134a)	0	0.6–0.94	3750	–
	R410A(0.5R32/0.5R125)	0	0.5	1890	–
	R428(0.775R125/0.2R134a/0.019R600a/0.006R290)	0	–	3500	–
	FX40(0.1R32/0.45R125/0.45R143)	0	0.6	3350	–
Zeotropic mixtures	R407A(0.2R32/0.4R125/4R134a)	0	0.14–0.45	1920	–
	R407B(0.1R32/R0.7R125/2R134a)	0	0.1–0.5	2560	–
	R407C(0.23R32/0.25R125/0.52R134a)	0	0.29–0.37	1610	–

(Continued)

Table 3.4 (Continued)

Group	Fluid	ODP	GWP (R11 = 1)	GWP (CO_2 = 1)	Atmospheric Lifetime (years)
	R417A(0.466R125/0.5R134a/ 0.034R600)	0	–	2300	–
	R422A(0.851R125/0.115R134a/ 0.034R600a)	0	–	3100	–
	R424(0.505R125/0.47R134a/ 0.009R600a/0.01R600/ 0.006R60)	0	–	2400	–
	R427A(0.15R32/0.25R125/ 0.1R143a/0.5R134a)	0	–	2100	–

cycles with evaporation temperature $t_0 = -15\,°C$ and condensation temperature $t_c = +30\,°C$.

Cycle data are available from different sources [7,9], or can be evaluated from suitable software such as REFPROP [16].

The selection of refrigerants in Table 3.5 has been made in order to present the overview of cycle data for historically used natural inorganic refrigerants such as R717, R744, R764 (which is not in use anymore), CFCs such as R11 or R12 and HCFCs such as R22, and mixture R502. Amongst newly used refrigerants HFCs R32 and R134a are presented as well as zeotropic mixtures of HFCs R404A, R407C, R410A, and azeotropic mixtures of HFCs R507. Finally, natural HCs R600a and R290, together with propylene R1270 are listed.

As it can be seen from data presented in Table 3.5, pressures in the system are temperature-dependent and are different for each particular refrigerant. Evaporation and condensation temperatures are closely coupled with corresponding pressure for single-component refrigerants, while for zeotropic mixtures temperature glide appears during the phase change at constant pressure.

Pressures influence design and thus equipment costs, but also the power consumption for compression and thus operational costs. Refrigerant transport properties, such as liquid and vapour density, viscosity, and thermal conductivity define heat transfer coefficients and consequently, temperature differences in heat exchangers thus directly influence pressures in the system as well as the necessary heat transfer surface of heat exchangers. Molecular mass or volumetric refrigerating capacity of some refrigerants influences application of certain compressor types. For example, NH_3 systems are not suitable for application of centrifugal compressor due to the low molecular mass of NH_3 The higher the volumetric refrigeration capacity is, the smaller compressor displacement can be, which results in smaller compressors for refrigerants with high volumetric refrigeration capacities. A good example is R744 which has the highest volumetric capacity.

Achievable efficiency of the entire process is due in great part to the refrigerant used. Effective energy consumption or COP is not equal to the one of the theoretical cycle. Isentropic efficiency η_{is} in Eqn (2.8) is also dependent on refrigerant properties.

Table 3.5 Parameters of −15/30 °C Cycle with Different Refrigerants

Refrigerant	p_0 (bar)	p_c (bar)	p_c/p_0 (–)	q_{0v} (kJ/m³)	COP (–)	t_2 (°C)	Safety Code
R717	2.362	11.672	4.942	2167.6	4.76	99.08	B2L
R744	22.90	72.10	3.149	7979.0	2.69	69.50	A1
R764	0.807	4.624	5.730	818.8	4.84	96.95	B1
R11	0.202	1.260	6.233	204.2	5.02	42.83	A1
R12	1.823	7.437	4.079	1273.4	4.70	37.81	A1
R22	2.962	11.919	4.024	2096.9	4.66	52.95	A1
R32	4.881	19.275	3.949	3420.0	4.52	68.54	A2L
R134a	1.639	7.702	4.698	1225.7	4.60	36.61	A1
R404A	3.610	14.283	3.956	2099.1	4.16	36.01	A1
R407C	2.632	13.591	5.164	1802.9	3.91	51.43	A1
R410A	4.800	18.893	3.936	3093.0	4.38	51.23	A1
R502	3.437	13.047	3.796	2079.5	4.39	37.07	A1
R507	3.773	14.600	3.870	2163.2	4.18	35.25	A1
R600a	0.891	4.047	4.545	663.8	4.71	32.66	A3
R290	2.916	10.790	3.700	1814.5	4.55	36.60	A3
R1270	3.630	13.050	3.595	2231.1	4.55	41.85	A3

Discharge temperature on the compressor outlet t_2 depends on refrigerant and systems pressures, and it must be limited in order to avoid deterioration of oil properties, or even the oil burnout. Behaviour of some refrigerants during the compression can result is no or low superheating of the vapour at the end of the compression (e.g. R134a which has low superheating, or R600a where final refrigerant state at the end of the compression can end in a saturated area unless proper superheating at the compressor inlet is provided). Systems with such refrigerants are not suitable for utilisation of the superheated part of vapour heat content in refrigeration cycles with heat recovery for sanitary water heating during the cooling operation [15].

Pressure drop within heat exchangers and in pipelines connecting refrigeration machine components are essential for system efficiency and are also dependent of refrigerant properties.

3.4 STRATEGY CONCERNING NON-ECOLOGICAL REFRIGERANTS

At the 15th session of the Conference of Parties to the United Nations Framework Convention on Climate Change (UNFCCC) in Copenhagen, Denmark, in 2009, climate change was underlined as one of the greatest challenges. Deep cuts in global emission and low-emission development strategy were recognised as crucial to combat the issue [17]. These calls to curb GHGs emissions are continuations of the 1997 Kyoto Protocol [3], which calls for the reduction of the emissions of, among others, CO_2 and two groups of refrigerants; HFCs and per-fluorocarbons. The production of these refrigerants was regulated even earlier, since 1987, under the Montreal Protocol [2]. Countries, trade associations and companies are increasingly adopting regulations and voluntary programs to minimise these releases and, as a result, minimise potential environmental effects while continuing to allow use of these refrigerants.

In response, more environmentally friendly refrigeration systems have been investigated in recent years [18–21]. Two aspects are of particular concern, namely the use of ecological (environmentally friendly) refrigerants and the energy consumption issue.

The thermodynamic and thermo-physical properties of refrigerants must be carefully analysed and taken into account during the conception and design of the cooling systems because they influence the energy performance of the system and have an environmental impact.

The normal evaporation temperatures of the primary refrigerants are presented in Figure 3.1, and the principal thermo-physical properties of some natural refrigerants are summarised in Table 3.3 [7,8]. Of these properties, the normal boiling point is most important because it is a direct indicator of the temperature at which a refrigerant can be used. The freezing point must be lower than any contemplated usage. The critical properties describe a material at the point where the distinction between liquid and gas is lost. The specific heat ratio should be low. Hence, lower discharge temperature can be expected, which will improve the compressor life. Molecular mass of the refrigerant affects the compressor size because the specific volume of the vapour is directly related to it. A low molecular mass refrigerant is preferred for the reciprocating refrigerant compressor. Reciprocating compressors are preferred for refrigerants that have high pressure and a small volume of vapour. Rotary compressors are used with refrigerants having low pressure and a large volume of suction vapour.

3.4.1 Pure Halogenated Refrigerants as Alternatives

The CFC refrigerants of R11 and R12 were substituted by simpler compound refrigerants R123 (HCFC) and R134a (HFC) with a reduced or even zero impact on the depletion of the ozone layer [22]. This alternative is attractive because the substitutes have similar properties (i.e. temperature, pressure) with the replaced refrigerants and the changes that occur directly on the existing installations are realised with a minimum investment.

Additionally, the substitution of R123 or R11 refrigerants with R22 or R134a, which have molecular masses lower by 50%, leads to reduced dimensions of the refrigeration equipment by 25–30% [23].

For other refrigerants, no simple compound fluids, R502 for example, could be replaced with a mixture of R115 (CFC) and R22 (HCFC) or in some cases only with R22, which is a fluid for temporary substitution. However, all these compounds are considered to be GHGs. As a response to these concerns, even more ecological refrigerants, mainly R1234yf [24] and natural refrigerants [25–27], particularly CO_2 and NH_3, have been proposed as substitutes.

The refrigerant R12, which could deplete the ozone layer, was replaced by R134a in the 1990s. However, R134a is still a GHG whose GWP is approximately 1300. During the Kyoto Protocol [3], R134a was already on the list of refrigerants with restricted use.

Previous studies have considered R152a [28] and the natural refrigerant CO_2 [27] as possible substitutes for R134a in automotive A/C systems. However, R152a is a flammable refrigerant, which must use a secondary loop when used in automotive A/C systems. The working pressure required for the CO_2 system is much higher than that

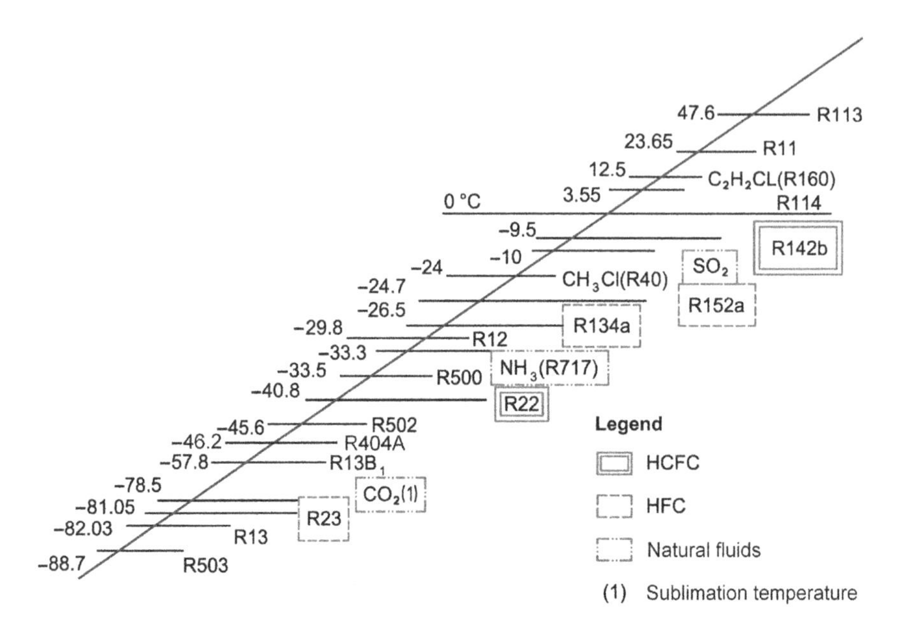

Figure 3.1 Evaporation temperatures of some refrigerants.

for the R134a system. Extremely high pressure leads to a significant change in the present system and thus a higher system cost. Hence, these two refrigerants are far from practical for use as substitutes of R134a in present-day automotive A/C systems.

Recently, R1234yf was proposed as an alternative of R134a in automotive A/C systems [18]. R1234yf has an ODP of 0, and its GWP is only 4. Thus, R1234yf satisfies the recent environmental requirements and polices quite well. This refrigerant has been classified as a very low flammable working fluid (A2L safety group [7]). In addition, the thermo-physical properties of R1234yf are quite similar to those of R134a. The main thermodynamic and thermo-physical properties of R1234yf are summarised in Table 3.6 and compared with those of R134a. The working pressure of the R1234yf system is very close to that of the R134a system under the working conditions of automotive A/C systems. Mathur [29] compared the performance of A/C systems using either R1234yf or R134a as working fluids and found that the performances of these two refrigerants in the same system are comparable under the same working conditions. Furthermore, R1234yf has a low toxicity that is similar to that of R134a. The use of a mini-channel heat exchanger is recommended in R1234yf systems because it can significantly reduce the refrigerant's charge [20].

3.4.2 Natural Refrigerants as Alternatives

Among the natural refrigerants, CO_2 (R744) seems to be the most promising. CO_2 has many excellent advantages in engineering applications, such as no toxicity, inflammability, high volumetric capacity (with a possibility to make the system compact), lower pressure ratio, superior heat transfer properties, complete compatibility with normal lubricants, easy availability, lower price and no recycling issues.

CO_2 was once widely used as a refrigerant, particularly in industrial and marine systems. After 1950, however, it was abandoned in favour of CFCs because the conventional CO_2 systems require high power consumption and lose capacity at high temperature. Since Lorentzen [27] proposed CO_2 as a possible natural refrigerant, a number of studies have been performed for different types of HP, A/C and refrigeration systems.

However, because the critical temperature of CO_2 (31.1 °C) is usually lower than the typical value of the heat rejection temperature of HP and A/C systems, the transcritical vapour-compression cycle, rather than a conventional one, is appropriate for the use of CO_2 for water heating and for comfort cooling and heating [30].

Applications of a transcritical CO_2 cycle in domestic water heating systems exhibited advantages over conventional systems with respect to power consumption and heating efficiency [31].

Various methods have been proposed to improve the energy efficiency of HP systems. One of these methods is to recover the power loss during expansion by replacing the conventional expansion valve with an expander; this idea was initially proposed to increase the efficiency of the CO_2 refrigeration system [27]. When applied to conventional R22 and R134a systems, the COP was reported to increase by up to 15% [32] and 12% [33], respectively. When applied to a transcritical CO_2 system, where the pressure difference between the suction and discharge lines is very high (in the range of 70 bars), the COP can increase by up to 50% [34].

Inagaki et al. [35] found that for a two-stage CO_2 A/C cycle, the capacity and COP are improved by 35% and 20%, respectively, at moderate ambient temperature and the capacity and COP are improved by 10% and 5%, respectively, at high ambient temperature conditions.

To improve the COP of the cycle, an ejector can be used instead of a throttling valve to recover some of the kinetic energy of the expansion process. Li and Groll [36], using a theoretical model, showed that the COP of a CO_2 cycle can be improved by more than 16% using an ejector.

CO_2 has been used in European supermarkets to a significant extent, but it is inherently inefficient in use compared with R22 except when used in tandem with another refrigerant. The working pressure required for the CO_2 system is much higher than that for the R134a system. High working pressure and high isothermal compression coefficients are some issues that should be considered. Subcritical CO_2 systems

Table 3.6 Main Properties of R1234yf and R134a

No.	Property	R1234yf	R134a
1	Chemical formula	$CF_3CF = CH_2$	$C_2H_2F_4$
2	Boiling point, t_{0n} (°C)	−29	−26
3	Critical temperature t_{cr} (°C)	95	102
4	Liquid density, ρ_l at 25 °C (kg/m^3)	1094	1207
5	Vapour density, ρ_v at 25 °C (kg/m^3)	37.6	32.4
6	ODP	0	0
7	GWP	4	1300

are less efficient than NH_3 systems, and transcritical systems are even less efficient. It will be difficult to justify the use of CO_2 as a general substitute for R22 except in cool and temperate regions.

CO_2 is not usually used in chillers, mostly due to the low energy efficiency of the process. CO_2 HPs for water heating started selling in Japan in 2001. They can heat domestic water up to 70–80 °C. The capacity of those chillers goes up to 100 kW. The Japanese have focused considerable attention on CO_2 HPs, such as the EcoCute water-to-water HP, the Unimo air-to-water HP, and the Sirocco water-to-air HP. High efficiency is an important benefit of such systems; they operate at a COP of about 4.0. If they are configured to provide space cooling in addition to hot water, the COP can be as high as 8.0. EcoCute is an ecologically-efficient electric HP water heater. The CO_2 HP water heater cycle is transcritical, operating at much higher temperatures and pressures than conventional subcritical cycles. In 2001, the first EcoCute HPs were sold. As of today, more than 3.5 million units have been installed in Japan, while yearly sales have been constantly increasing, reaching more than 550,000 units sold per year [37,38]. European CO_2 HP manufacturers are also emerging, with different companies adding CO_2 HPs to their product ranges [39].

Table 3.4 shows that NH_3 (R717) is the only substitute for R22 that has zero GWP. Therefore, between the natural fluids, the NH_3 is the best substitute for R22 that has favourable thermodynamic properties (Table 3.3), a high heat transfer coefficient (3–4 times superior to R22, in accordance with Table 3.7) and a COP similarly good for many applications, especially industrial ones, with great cooling potentials [26]. NH_3 is cheap and ecological (ODP = 0, GWP = 0). However, NH_3 has a high toxicity (class B), but has a characteristic, sharp smell which makes a warning possible below concentrations of 3 mg/m^3 a NH_3 in air. Also, NH_3 has mild flammability. This refrigerant is under the 2L flammability subclass [7]. However, its ignition energy is 50 times higher than that of natural gas and NH_3 will not burn without a supporting flame. The market opportunity produced by R22 phase-out should not be missed by producers of NH_3 chillers. The major obstacles are legal demands in some countries as well as high initial costs because of present production in small series. NH_3 interest renewed in Europe and especially in northern Europe.

Table 3.8 presents the contribution of some refrigerants to atmosphere warming for an evaporation temperature $t_0 = -20$ °C, a condensation temperature $t_c = 35$ °C, and an operating time of 15 years. Because the drive energy of system is reduced for R717, this refrigerant has a TEWI index lower than that of other working fluids. The NH_3 also has a high critical temperature that tends to make systems using it more efficient than systems using other refrigerants.

Table 3.7 Heat Transfer Coefficients, in W/m^2 K, for R717 and R22

No.	Specifications	R717	R22
1	Outside tube condensation	7500 – 11,000	1700 – 2800
2	Inside tube condensation	4200 – 8500	1400 – 2000
3	Outside tube evaporation	2300 – 4500	1400 – 2000
4	Inside tube evaporation	3100 – 5000	1500 – 2800

HCs like propane (R290), propylene (R1270) or isobutene (R600a) have been used in refrigeration systems all over the world for many years. HCs are colourless and nearly odourless gases that liquefy under pressure, and have neither ODP (0) nor significant direct GWP (<3). Thanks to their thermodynamic properties, HCs make particularly energy efficient refrigerants and are commonly used in small systems with low refrigerant charges. However, the HCs have a high flammability and are under the A3 safety code [7]. In spite of the flammability of HCs, many companies especially in Europe and Asia use HCs as refrigerants without any hazard to consumers [40]. R600a and isobutane mixtures have displaced R12 and later R134a and now dominate in domestic refrigerators in Europe, but not in United States [9]. Typical refrigerator sizes are larger in the United States than in Europe, but are more comparable to those in Japan and Korea where isobutane usage is also increasing in refrigerators and vending machines.

3.4.3 Refrigerant Mixtures as Alternatives

Very limited pure fluids have suitable properties to provide alternatives to the existing halogenated refrigerants. The mixing of two or more refrigerants provides an opportunity to adjust the properties, which are most desirable. The three categories of mixtures used in A/C and refrigeration applications are azeotropes, near azeotropes (quasi-azeotropes) and zeotropes [41].

An azeotropic mixture of the substances cannot be separated into its components by simple distillation. Azeotropic mixtures have boiling points that are lower than either of their constituents. An azeotropic mixture maintains a constant boiling point and acts as a single substance in both liquid and vapour state. Azeotropic refrigerant mixtures are used in low temperature refrigeration applications.

The objective with near azeotropic mixtures is to extend the range of refrigerant alternatives beyond single compounds. Near azeotropes have most of the same attributes as azeotropes and provide much wider selection possibilities. However, near azeotropic mixtures may alter their composition and properties under leakage conditions.

Zeotropic refrigerant mixtures are blends of two or more refrigerants that deviate from perfect mixtures. A zeotropic mixture does not behave like a single substance when it changes state. Instead, it evaporates and condenses between two temperatures (temperature glide). The phase change characteristics of the zeotropic refrigerant

Table 3.8 Refrigerants Contribution to Atmosphere Warming

Refrigerant	Direct Effect		Indirect Effect	
	Operating Leak (kg CO_2)	Fluid Recovery Leak (kg CO_2)	Drive Energy Generation (kg CO_2)	TEWI (kg CO_2)
R22	1,033,500	68,900	1,805,400	2,907,800
R134a	911,625	60,775	1,884,150	2,856,550
R407C	999,352	66,623	2,104,650	3,170,625
R410A	1,049,555	69,970	1,962,900	3,082,425
R717	0	0	1,457,550	1,457,550

mixture (boiling and condensation) are non-isothermal. Zeotropic substances have greater potential for improvements in energy efficiency and capacity modulation. However, the major drawback of the zeotropic refrigerant mixture is the preferential leakage of more volatile components leading to change in mixture composition.

Figure 3.2 illustrates a strategy concerning the refrigerants.

During the last decade, many experimental and theoretical investigations have been reported with new refrigerant mixtures used in different applications, such as, HP, A/C, domestic and commercial refrigeration, chiller and automobile A/C.

It is very important not only to have a refrigeration system using an environmentally-friendly refrigerant, but also to have one with good energy efficiency. One of the ways to improve the energy efficiency of refrigeration systems is to recover the power lost during expansion using an expander [27]. The expander acts as a compressor operating in reverse, which improves the coefficient of performance of the system in two ways [42]: (i) by increasing the cooling capacity through performing a near-isentropic expansion, hence reducing the enthalpy of the refrigerant at the evaporator inlet, and (ii) by recovering the expansion energy, hence reducing the external electrical power requirement of the compressor. Expanders are especially attractive for the R404A and the transcritical CO_2 systems [21].

Many HC refrigerant mixtures were developed to substitute the halogenated refrigerants. HC mixtures are miscible with both mineral oil and synthetic lubricants. Hence, HC mixtures can be used as substitutes without changing the lubricant in the existing systems using HFC and HCFC refrigerants.

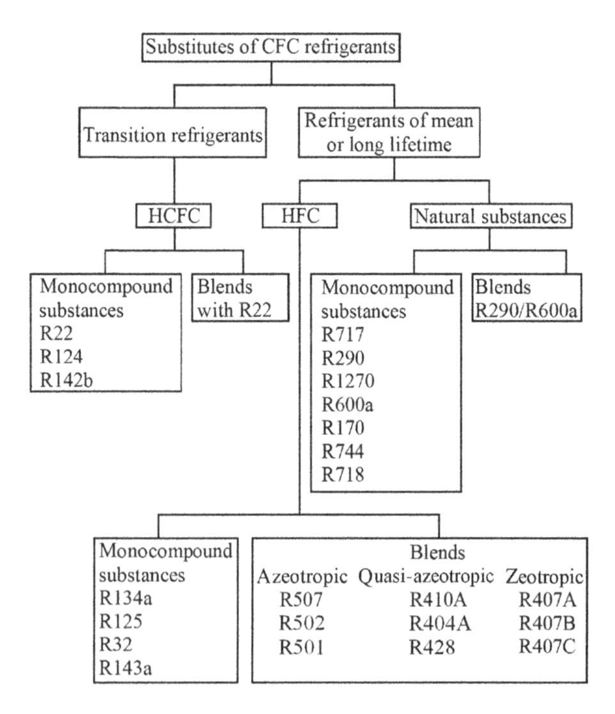

Figure 3.2 Strategy concerning the refrigerants.

The HFC mixtures such as R404A, R407C and R410A are reported as potential alternatives to R22 in HP, A/C and refrigeration systems [43]. However, HFC mixtures are not miscible with mineral oil, which is used as a lubricant in CFC and HCFC systems. HFC mixtures require a synthetic lubricant, like polyolester. Hence, a major modification is required for HFC mixtures to retrofit in HCFC systems.

The HCFC mixtures are considered as interim alternatives due to their ODP. R123 is an HCFC refrigerant that has a very low value of ODP, with lower GWP of 120.

Cox et al. [44] developed a new NH_3-based azeotropic mixture (R717/R170). It has been reported that this mixture has lower compressor discharge temperature, which favours system reliability and improves the cycle efficiency. The mixture has good miscibility with mineral oil, thereby reducing the usage of highly hygroscopic synthetic oils.

The suitability of new refrigerant mixtures in the existing refrigeration system requires further research in the following main areas:

1. Reliability of refrigerant compressors working with environment-friendly alternatives.
2. Development of a new user-friendly lubricant is necessary to replace the existing synthetic lubricant.
3. Development of a new method for heat exchanger design is required to accommodate the non-linear variation of new refrigerant mixtures during phase change.
4. The environmental properties, flammability and safety issues of new refrigerant mixtures.
5. Thermoeconomic optimisation of vapour-compression refrigeration systems working with the new refrigerant mixture.

3.5 CONCLUSIONS

This chapter contains a good amount of information regarding ecological refrigerant trends. Scientific research based on monocompound substances or mixtures will lead to the discovery of adequate substitutes for cooling applications that will not only be ecological ($ODP = 0$, reduced GWP), non-flammable and non-poisonous, but also have favourable thermodynamic properties.

A possible solution is the use of inorganic refrigerants (i.e. NH_3, CO_2) and HC refrigerants (e.g. propane, isobutene, ethylene, propylene) for industrial applications and A/C systems. Because the HC refrigerant represents a high risk of flammability and explosion, these substances will not often be used as refrigerants, compared with CO_2 or NH_3 usage. Another advantage of these two substances is that they were used as refrigerants for a long time.

The European Partnership for Energy and Environment considers the HFC refrigerants as the best alternative for the refrigerants CFC and HCFC in most of the applications. The HFC refrigerants allow for energetically efficient applications, offering significant benefits compared with the existing alternatives. On average, more than 80% of the gases that contribute to the greenhouse effect and are used in cooling equipment come from indirect emissions. The high energy efficiency of the use of HFC refrigerants balances a great measure of their GWP.

The new refrigerant generation must offer high efficiency or the change to address low GWP will backfire with increased rather than decreased net GHG emissions.

There are cases with more options for an alternative refrigerant, for which the problem is to choose the economical variant. The replacement of some refrigerants with other non-polluting influences the operating conditions of the cooling systems, due to a rapid degradation of components made from elastomers [45] or plastic materials [18], or is necessary to replace mineral oils with some other oils adequate to the new refrigerants. The problems of materials endurance and compatibility can be solved only by performing many tests, but the estimation of the energetic performances and expenses that result from the modification of the operational characteristics when replacing the refrigerant can be solved using numerical modelling [20].

The influence of refrigerant properties and refrigeration system design is significant and those properties influence the design of HVAC systems which also contain refrigeration subsystems. A new concept in the implementation of refrigeration systems is now imposed: it must be tightly constructed, with refrigerants having a reduced atmospheric warming potential but with a performance that is as energetically efficient as possible. Given the scarcity of viable options, future refrigerant selections warrant collective consideration of all environmental issues together with integrated assessments.

The use of ecological refrigerants plays a crucial role for reducing the environmental impacts of the halogenated refrigerants and to protect the environment.

REFERENCES

[1] ARAP web site. Alliance for Responsible Atmospheric Policy. <http://www.arap.org>.

[2] UNEP. Montreal protocol on substances that deplete the ozone layer. New York, NY: United Nations Environment Program; 1987.

[3] GECR. Kyoto protocol to the United Nations framework conservation on climate change. Global Environmental Change Report. New York, NY; 1997.

[4] Richard LP. CFC phase out; have we met the challenge. J Fluorine Chem 2002;114:237–50.

[5] Johnson E. Global warming from HFC. Environ Impact Assess Rev 1998;18:485–92.

[6] ASHRAE Standard 34. Designation and safety classification of refrigerants. Atlanta, GA: American Society of Heating, Refrigerating and Air-Conditioning; 2007.

[7] ASHRAE Handbook. Fundamentals. Atlanta, GA: American Society of Heating, Refrigerating and Air-Conditioning; 2013.

[8] Sarbu I. A review on substitution strategy of non-ecological refrigerants from vapour compression-based refrigeration, air-conditioning and heat pump systems. Int J Refrig 2014;46(10):123–41.

[9] Calm JM, Hourahan GC. Refrigerant data summery. Eng Syst 2001;18:74–88.

[10] Palm B. Hydrocarbons as refrigerants in small heat pump and refrigeration systems – a review. Int J Refrig 2008;31:552–63.

[11] Formeglia M, Bertucco A, Brunis S. Perturbed hard sphere chain equation of state for applications to hydro fluorocarbons, hydrocarbons and their mixtures. Chem Eng Sci 1998;53:3117–28.

[12] Sarbu I, Sebarchievici C. Heat pumps. Timisoara: Polytechnic Publishing House; 2010 [in Romanian]

[13] Dragos GV, Dragos R. Use of CO_2 in non-polluting refrigerating systems. In: Proceedings of the 28th conference "Modern Science and Energy". Cluj-Napoca, Romania: Risoprint; 2009. p. 149–57.

[14] COM643, Proposal for a regulation of the European Parliament and of the Council on fluorinated greenhouse gases; 2012.

[15] Pavkovic B. Refrigerant – properties and air-conditioning applications. Rehva J 2013;50(5):7–11.

[16] Lemmon EW, Huber ML, McLinden MO. REFPROP Reference fluid thermodynamic and transport properties. NIST Standard Reference Database 23 Version 9.1, US Secretary of Commerce; 2013.

[17] UNFCCC. United Nations Framework Convention on Climate Change. Warsaw, Poland; 2013.

[18] Minor B, Spatz M. HFO-1234yf low GWP refrigerant update. In: International refrigeration and air-conditioning conference at Purdue, West Lafayette, IN, USA, Paper no. 1349; 2008.
[19] Aprea C, Greco A, Maiorino A. An experimental evaluation of the greenhouse effect in the substitution of R134a with CO_2. Energy 2012;45:753–61.
[20] Zhao Y, Liang Y, Sun Y, Chen J. Development of a mini-channel evaporator model using R1234yf as working fluid. Int J Refrig 2012;35:2166–78.
[21] Subiantoro A, Ooi KT. Economic analysis of the application of expanders in medium scale air-conditioners with conventional refrigerants, R1234yf and CO_2. Int J Refrig 2013;36:1472–82.
[22] Agrawal AB, Shrivastava V. Retrofitting of vapour compression refrigeration trainer by an eco-friendly refrigerant. Indian J Sci Technol 2010;3(4):6837–46.
[23] Wright B. Environment forum. New York, NY: Carrier Air Conditioning Company; 1992.
[24] Honeywell <http://www.1234facts.com>.
[25] Hwang Y, Ohadi M, Radermacher R. Natural refrigerants. Mech Eng 1998;120:96–9.
[26] Lorentzen G. Ammonia: an excellent alternative. Int J Refrig 1988;11:248–52.
[27] Lorentzen G. Revival of carbon dioxide as a refrigerant. Int J Refrig 1994;17(5):292–301.
[28] Ghodbane M. An investigation of R152a and hydrocarbon refrigerants in mobile air-conditioning. In: Proceedings of the international congress and exposition, Paper no, 1999-01-0874; 1999.
[29] Mathur GD. Experimental investigation of AS system performance with HFO-1234yf as the working fluid. SAE paper 2010-01-1207; 2010.
[30] Liao SM, Zhao TS, Jakobsen A. A correlation of optimal heat rejection pressure in trans-critical carbon dioxide cycles. Appl Therm Eng 2000;20(9):831–41.
[31] Neksa P. CO_2 heat pump systems. Int J Refrig 2002;25:421–7.
[32] Robinson DM, Groll EA. Efficiencies of trans-critical CO_2 cycles with and without an expansion turbine. Int J Refrig 1998;21:577–89.
[33] Goncalves VDA, Parise JAR. A study on the reduction of throttling losses in automotive air-conditioning systems through expansion work recovery. In: International refrigeration and air-conditioning conference at Purdue; 2008. p. 2416.
[34] Fukuta M, Yanagisawa T, Nakaya S, Ogi Y. Performance and characteristics of compressor/expander combination for CO_2 cycle. In: The seventh IIR-Gustav Lorentzen conference on natural working fluids. Trondheim, Norway; 2006. p. 052.
[35] Inagaki M, Sassaya H, Osakli Y. Pointing to the future two stage CO_2 compression, heat transfer issues in natural refrigerants. International Institute of Refrigeration; 1997. p. 131–40
[36] Li D, Groll EA. Trans-critical CO_2 refrigeration cycle with ejector-expansion device. Int J Refrig 2005;28:766–73.
[37] Gillaux S. Inside view into the Japanese hat pump market. Rehva J 2012;49(5):55–6.
[38] Goto H, Goto M, Sueyoshi T. Consumer choice on ecologically efficient water heaters: marketing strategy and policy implications in Japan. Energy Econ 2011;33:195–208.
[39] Maratou A, Lira JT, Jia H, Masson N. CO_2 heat pumps in Europe market dynamics & legislative opportunities. Rehva J 2012;49(5):50–2.
[40] Fatouh M, Kafafy M. Experimental evaluation of a domestic refrigerator working with LPG. Appl Therm Eng 2006;26:1593–603.
[41] Didion DA, Bivens DB. Role of refrigerant mixtures as alternatives to CFCs. Int J Refrig 1990;13:163–75.
[42] Nickl J, Will G, Quack H, Kraus WE. Integration of a three-stage expander into a CO_2 refrigeration system. Int J Refrig 2005;28:1219–24.
[43] Calm JM, Domanski PA. R22 replacement status. ASHRAE J 2004;46:29–39.
[44] Cox N, Mazur V, Colbourne D. The development of azeotropic ammonia refrigerant blends for industrial process applications. In: International conference on ammonia refrigeration technology. Ohrid; 2009.
[45] Eiseman BJ. Effect on elastomers of Freon compounds and other halo-hydrocarbons. Refrig Eng 1949;12:1171.

CHAPTER 4

Types of Compressors and Heat Pumps

4.1 REFRIGERATION COMPRESSORS

A refrigeration compressor is a mechanical device that increases the pressure of a gas (refrigerant vapour) by reducing its volume. Heat pump (HP) and air-conditioner (A/C) equipment use compressors to move heat in refrigerant cycle.

The main types of HP compressors according to European Heat Pump Association are illustrated in Figure 4.1.

Compressors used in refrigeration systems are often described as being hermetic, open or semi-hermetic.

Typically in hermetic and most semi-hermetic compressors, the compressor and motor driving the compressor are integrated and operate within the refrigerant system. The motor is hermetic and is designed to operate, and be cooled by, the refrigerant being compressed. The disadvantage of hermetic compressors is that the motor drive cannot be repaired or maintained, and the entire compressor must be removed if a motor fails.

An open compressor has a motor drive which is outside of the refrigeration system, and provides drive to the compressor by means of an input shaft with suitable gland seals. Open compressor motors are typically air-cooled and can be fairly easily exchanged or repaired without degassing of the refrigeration system. The disadvantage of this type of compressor is the potential failure of the shaft seals, leading to loss of refrigerant.

Open motor compressors are generally easier to cool (using ambient air) and therefore tend to be simpler in design and more reliable, especially in high-pressure applications where compressed gas temperatures can be very high. However, the use of liquid injection for additional cooling can generally overcome this issue in most hermetic motor compressors.

4.1.1 Reciprocating Compressors

A reciprocating compressor or piston compressor is a positive-displacement compressor that uses pistons driven by a crankshaft to deliver gases at high pressure [1].

The intake gas enters the suction manifold, then flows into the compression cylinder where it gets compressed by a piston driven in a reciprocating motion via a crankshaft, and is then discharged.

Ground-Source Heat Pumps. DOI: http://dx.doi.org/10.1016/B978-0-12-804220-5.00004-7

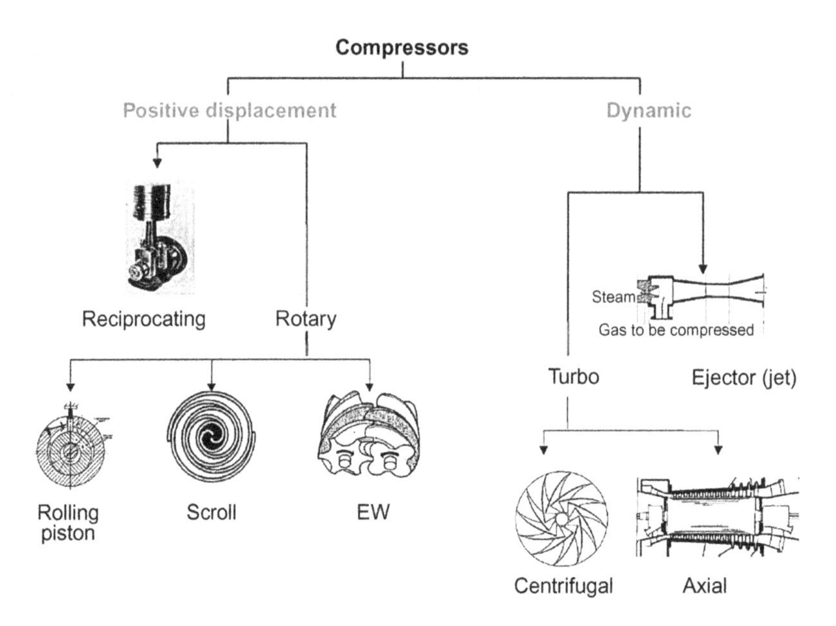

Figure 4.1 Types of heat pump compressors.

The reciprocating compressor has the following main characteristic parameters:

- theoretical vapour flow rate (displacement) D_0, in m^3/s:

$$D_0 = \frac{\pi d^2}{4}\frac{1}{60} sNn_k \tag{4.1}$$

where d is the cylinders' diameter, in m; N is the number of cylinders; s is the piston path, in m; and n_k is the compressor speed, in rev/min.

- real volumetric flow rate D, in m^3/s:

$$D = \lambda_k D_0 \tag{4.2}$$

where λ_k is the volumetric efficiency of compressor

- vapour mass flow rate m, in kg/s:

$$m = \frac{Q_{HP}}{q_c} = \frac{D}{v_1} = \frac{\lambda_k}{v_1} D_0 \tag{4.3}$$

where Q_{HP} is the HP capacity, in kW; q_c is the specific heat load at condensation, in kJ/kg; and v_1 is the specific volume of absorbed vapour, in m^3/kg.

- thermal power of the compressor Q_k, in kW:

$$Q_k = \lambda_k \frac{q_c}{v_1} D_0 \tag{4.4}$$

- pressure ratio H_p:

$$H_p = \frac{p_d}{p_s} \tag{4.5}$$

where p_d is the discharge pressure and p_s is the suction pressure.

- power consumed by the compressor P_k, in kW:

$$P_k = mw \tag{4.6}$$

where w is the specific compression work, in kJ/kg, which can be isothermal, adiabatic or polytrophic.

- indicated power P_i, in kW:

$$P_i = \frac{P_k}{\eta_i} = \frac{Q_{HP}}{COP_{hp}\eta_i} \tag{4.7}$$

where η_i is the indicated efficiency of the compressor ($\eta_i = T_0/T_c$).

- electric motor power of the compressor P_e, in kW:

$$P_e = \frac{Q_{HP}}{\eta_i \eta_{em} COP_{hp}} \tag{4.8}$$

where η_{em} is the electromechanical efficiency of the motor-compressor system.

4.1.2 Rotary Screw Compressors

Rotary screw compressors are also positive-displacement compressors. Two meshing screw-rotors rotate in opposite directions, trapping refrigerant vapour, and reducing the volume of the refrigerant along the rotors to the discharge point.

Compressor operation. Rotary screw compressors use two meshing helical screws, known as rotors, to compress the gas. In a dry running rotary screw compressor, timing gears ensure that the male and female rotors maintain precise alignment. In an oil-flooded rotary screw compressor, lubricating oil bridges the space between the rotors, both providing a hydraulic seal and transferring mechanical energy between the driving and driven rotor. Gas enters at the suction side and moves through the threads as the screws rotate (Figure 4.2a). The meshing rotors force the gas through the compressor (Figure 4.2b), and the gas exits at the end of the screws (Figure 4.2c).

The effectiveness of this mechanism is dependent on precisely fitting clearances between the helical rotors, and between the rotors and the chamber for sealing of the compression cavities.

The pressure ratio H_p is defined by:

$$H_p = \frac{p_d}{p_s} \tag{4.9}$$

where p_d is the discharge pressure and p_s is the suction pressure.

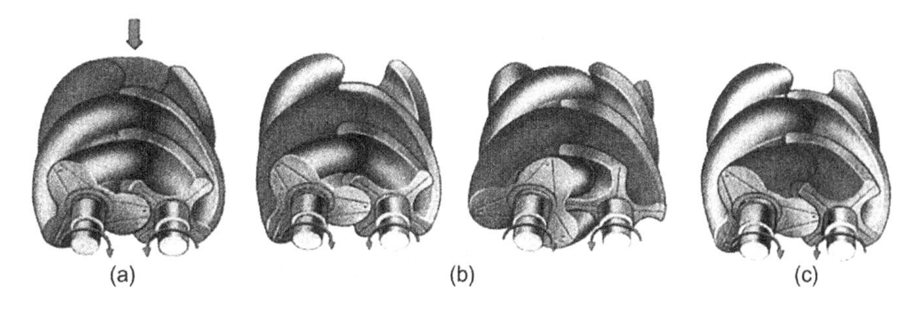

Figure 4.2 Phases of compression process.

Characteristic parameters. The theoretical flow rate D_0, in m^3/min, of the compressor is expressed as follows:

$$D_0 = \frac{n_k V_0}{1000} \tag{4.10}$$

where V_0 is the vapour volume transported to a main rotor rotation, in dm^3/rev, and n_k is the rotational speed of the main compressor shaft, in rev/min.

The real volumetric flow rate D of the compressor is smaller than D_0 due to volumetric losses:

$$D = \lambda_k D_0 \tag{4.11}$$

where λ_k is the volumetric efficiency of the compressor.

The power consumed by the compressor P_k, in kW, is given by the equation:

$$P_k = \frac{1}{60}\frac{D_0}{v_1}(i_2 - i_1) \tag{4.12}$$

where v_1 is the specific volume, in m^3/kg, of the refrigerant vapour in the suction state and i_1 and i_2 are the corresponding enthalpies, in kJ/kg, of the refrigerant for the aspiration and discharge conditions.

The final compression temperature t_2, in °C, for an oil-free compressor can be calculated using the equation:

$$t_2 = t_1 + \Delta t \tag{4.13}$$

in which:

$$\Delta t = \frac{T_1}{\lambda_k}\left[\left(\frac{p_d}{p_s}\right)^{\frac{\chi-1}{\chi}} - 1\right] \tag{4.14}$$

t_1 is the suction temperature, in °C; T_1 is the absolute suction temperature, in K; and χ is the isentropic index (approximately 1.4 for air and 1.2 for other gas).

The final temperature of the compressed working fluid can reach a maximum of 250 °C. This temperature corresponds to pressure ratio values of 4.5 and 7 for air and other gases, respectively.

In an oil-injected rotary screw compressor, oil is injected into the compression cavities to aid sealing and provide a cooling sink for the gas charge. The final compression temperature of the vapour is lower than 90 °C, and the pressure ratio can be $H_k \geq 21$.

Figure 4.3 shows the characteristic diagram of a screw compressor.

The thermal power Q_k, in kW, of a compressor in imposed running conditions is given by:

$$Q_k = \frac{1}{60}\lambda_k \frac{q_c}{v_1} D_0 \tag{4.15}$$

where q_c is the specific heat load at condensation, in kJ/kg.

The evaporator cooling power can be determined using the equation:

$$Q_0 = m q_0 = Q_k \frac{q_0}{q_c} = Q_k\left(1 - \frac{1}{\mathrm{COP_{hp}}}\right) \tag{4.16}$$

where q_0 is the specific cooling power of the working fluid, in kJ/kg.

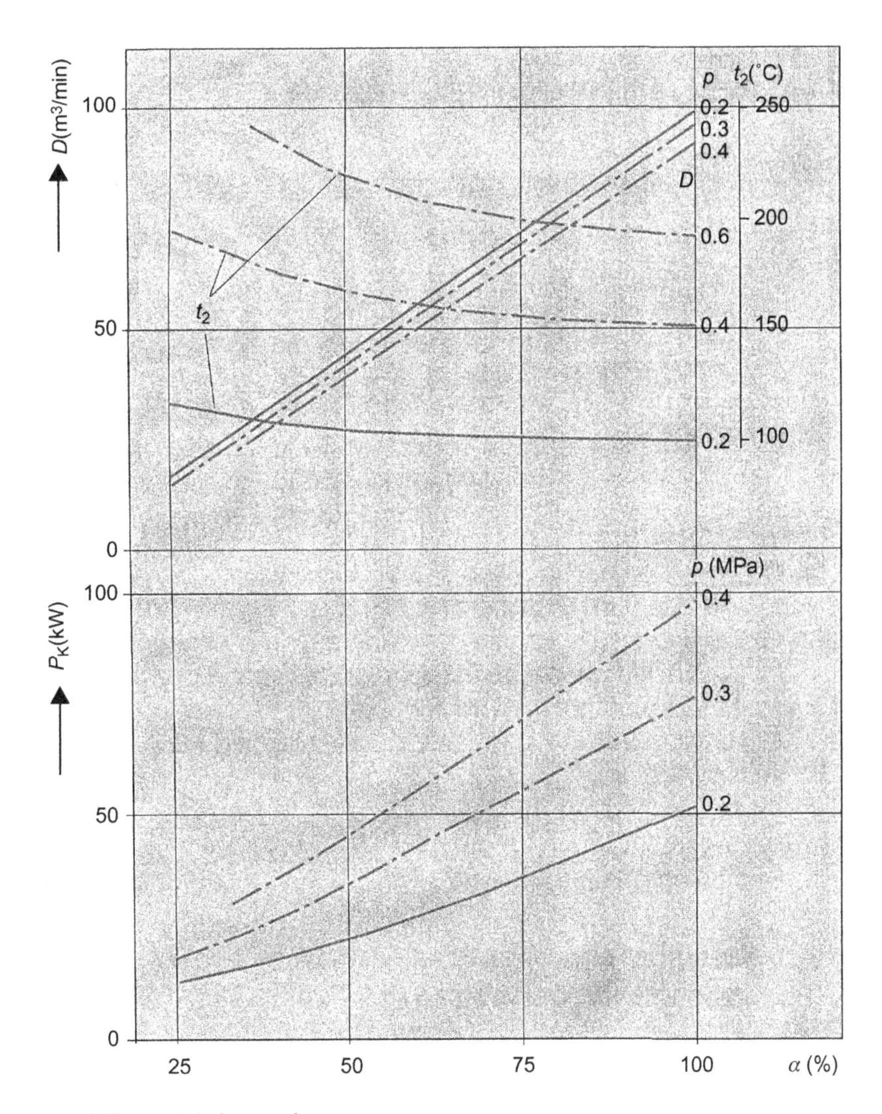

Figure 4.3 Characteristic diagram of a rotary screw compressor.

The coefficient of performance of the compressor ε_k is defined as:

$$\varepsilon_k = \frac{Q_0}{P_k} \tag{4.17}$$

In variable operational conditions, the rotary screw compressor is characterised by a variable absorbed vapour flow rate and compression power, that together achieve variable cooling powers. Figure 4.4a shows the cooling and compression power diagrams at different operational conditions for a screw compressor ($n_k = 3000$ rev/min, $D_0 = 816$ m^3/h) operating with R-22. Figure 4.4b shows the variation diagram of the efficiency ε_k.

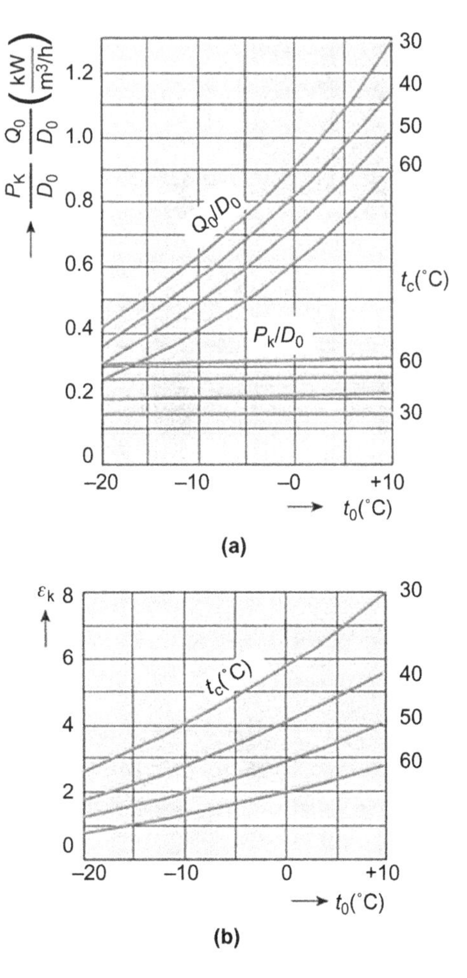

Figure 4.4 Operational characteristics variation of a screw compressor.

Large capacity HPs used in the heating/cooling of manufacturing facilities, office buildings, administrative buildings and hotels are equipped with screw compressors which operate alone or in parallel.

4.1.3 Centrifugal Compressors

Centrifugal compressors, sometimes termed radial compressors, are a sub-class of dynamic axisymmetric work-absorbing turbomachinery.

The idealised compressive dynamic turbo-machine achieves a pressure rise by adding kinetic energy/velocity to a continuous flow of fluid through the rotor or impeller. This kinetic energy is then converted to an increase in potential energy/static pressure by slowing the flow through a diffuser. The pressure rise in the impeller is almost equal to the rise in the diffuser section, in most cases.

In the case of where flow simply passes through a straight pipe to enter a centrifugal compressor; the flow is straight, uniform and has no vorticity. As the flow continues to pass into and through the centrifugal impeller, the impeller forces the flow to spin faster and faster. According to a form of Euler's fluid dynamics equation, known as the 'pump and turbine equation', the energy input to the fluid is proportional to the flow's local spinning velocity multiplied by the local impeller's tangential velocity.

A simple centrifugal compressor has four components: inlet, impeller/rotor, diffuser and collector [2].

The *inlet* to a centrifugal compressor is typically a simple pipe. It may include features such as a valve, stationary vanes/airfoils (used to help swirl the flow) and both pressure and temperature instrumentation. All of these additional devices have important uses in the control of the centrifugal compressor.

The *centrifugal impeller* is the key component that makes a compressor centrifugal. The impeller contains a rotating set of vanes (or blades) that gradually raises the energy of the working gas. This is identical to an axial compressor with the exception that the gases can reach higher velocities and energy levels through the impeller's increasing radius.

The *diffuser* is the next key component to the simple centrifugal compressor. Downstream of the impeller in the flow path, it is the diffuser's responsibility to convert the kinetic energy (high velocity) of the gas into pressure by gradually slowing (diffusing) the gas velocity. Diffusers can be vane-less, vanes or an alternating combination.

The *collector* of a centrifugal compressor can take many shapes and forms. When the diffuser discharges into a large empty chamber, the collector may be termed a *plenum*. When the diffuser discharges into a device that looks somewhat like a snail shell, bull's horn or a French horn, the collector is likely to be termed a *volute* or *scroll*. A collector's purpose is to gather the flow from the diffuser discharge annulus and deliver this flow to a downstream pipe.

Due to the wide variety of vapour compression-based cycles and the wide variety of refrigerants, centrifugal compressors are employed in a wide range of sizes and configurations.

4.1.4 Scroll Compressors

A scroll compressor (also called *spiral compressor*), is a device for compressing air or refrigerant. Many residential central HP and A/C systems employ a scroll compressor instead of the more traditional rotary and reciprocating compressors.

A scroll compressor uses two interleaving scrolls to pump, compress or pressurise fluids such as liquids and gases [3]. The vane geometry may be involutes, Archimedean spiral or hybrid curves. Often, one of the scrolls is fixed, while the other orbits eccentrically without rotating, thereby trapping and pumping or compressing pockets of fluid between the scrolls.

The compression process occurs over approximately two to two and one-half rotations of the crankshaft, compared with one rotation for rotary compressors, and one-half rotation for reciprocating compressors. The scroll discharge and suction processes occur for a full rotation, compared to less than a half-rotation for the reciprocating suction process, and less than a quarter-rotation for the reciprocating discharge process. Reciprocating compressors have multiple cylinders (typically, anywhere from two to six), while scroll compressors only have one compression element.

Scroll compressors never have a suction valve, but depending on their application may or may not have a discharge valve. The use of a dynamic discharge valve is more prominent in high-pressure ratio applications, typical of refrigeration. The use of a dynamic discharge valve improves scroll compressor efficiency over a wide range of operating conditions, when the operating pressure ratio is well above the built-in pressure ratio of the compressors.

The isentropic efficiency of scroll compressors is slightly higher than that of a typical reciprocating compressor when the compressor is designed to operate near one selected rating point. The scroll compressors are more efficient in this case because they do not have a dynamic discharge valve that introduces additional throttling losses. However, the efficiency of a scroll compressor that does not have a discharge valve begins to decrease when compared with the reciprocating compressor at higher-pressure ratio operation.

The scroll compression process is nearly 100% volumetrically efficient in pumping the trapped fluid. The suction process creates its own volume, separate from the compression and discharge processes further inside. By comparison, reciprocating compressors leave a small amount of compressed gas in the cylinder, because it is not practical for the piston to touch the head or valve plate. The remnant gas from the last cycle then occupies space intended for suction gas. The reduction in capacity (i.e. volumetric efficiency, λ_k) depends on the suction and discharge pressures with greater reductions occurring at higher ratios of discharge to suction pressures.

Until recently, a powered scroll compressor could only operate at full capacity. Modulation of the capacity was accomplished outside the scroll set. In order to achieve partial-loads, engineers would bypass refrigerant from intermediate compression pocket back to suction, vary motor speed, or provide multiple compressors and stage them on and off in sequence.

Recently, scroll compressors have been manufactured that provide partial-load capacity within a single compressor. These compressors change capacity while running. While scroll compressors can also rely on vapour injection to vary the capacity, their vapour injection operation is not as efficient as that of reciprocating compressors. This inefficiency is caused by the continuously changing volume of the scroll compressor's compression pocket during the vapour injection process. As the volume is continuously in flux, the pressure within the compression pocket is also continuously changing which adds inefficiency to the vapour injection process.

Some of the best compressors with efficiencies of up to 60% of Carnot's theoretical limit are produced by Danfoss, Copeland, York, Trane, Embraco and Bristol compressor manufacturers.

4.2 DESCRIPTION OF HEAT PUMP TYPES

HPs are classified by (i) the heat source and sink; (ii) the heating and cooling distribution fluids; and (iii) the thermodynamic cycle. The following classifications can be made according to:

- function: heating, cooling, domestic hot-water (DHW) heating, ventilation, drying, heat recovery etc.
- heat source: ground, ground-water, water, air, exhaust air etc.
- heat source (intermediate fluid)-heat distribution: air-to-air, air-to-water, water-to-water, antifreeze (brine)-to-water, direct expansion-to-water etc.

4.2.1 Air-to-Air Heat Pumps

Air-to-air HPs are the most common and are particularly suitable for factory-built unitary HPs. These HPs are also found in controlled dwelling ventilation applications to enable an increase in the heat recovery from the exhaust air, and can even allow for the cooling of selected rooms. For these applications, various units are used. Air-to-air HPs have a full-hermetic compressor, finned heat exchangers for the evaporator and condenser, and an expansion valve as well as the necessary safety mechanisms.

As the outdoor air temperature decreases the heat demand increases and the HP capacity substantially decreases due to the efficiency reduction.

4.2.2 Water-to-Air Heat Pumps

Water-to-air HPs rely on water as the heat source and sink, and use air to transmit heat to or from the conditioned space. They include:

- ground-water heat pumps (QWHPs), which use ground-water from wells as a heat source and/or sink
- surface-water heat pumps (SWHPs), which use surface water from a lake, pond or stream as a heat source or sink
- solar-assisted HPs, which rely on low-temperature solar energy as the heat source.

4.2.3 Air-to-Water Heat Pumps

Air-to-water HPs [4] use outdoor air as their heat source and are mostly operated in bivalent heating systems, as well as for cooling, heat recovery and DHW production. The indoor unit contains the substantial components and is fitted indoors, protected from weather and freezing temperatures. The outdoor unit is connected to the indoor unit via refrigeration lines. Through the elimination of air ducts, extremely quiet, energy-efficient fans are made possible.

Generally, a fully-hermetic compressor (piston or scroll) with built-in, internal overload protection is used in these HPs. Stainless steel flat plate heat exchangers are used for the condenser. Expansion valve, weather-dependent defrost mechanism – preferably hot gas. Copper-tube aluminium finned evaporator. For quiet operation, an axial fan with low speed should be used. Electrical components and the controller are either integrated or externally mounted, depending on the manufacturer and model. Control of the heating system is commonly integrated.

The heat demand of a building depends on the climate zone in which it is located. In temperate climate conditions, such as in Romania, the heat demand Q_{inc} evolves from the minimum values in the provisional seasons (spring and autumn) to the maximum value in the cold season (Figure 4.5). The annual number of hours with the minimum outdoor air temperature represents approximately 10–15% of the total time for heating, which is why the selection of an HP to cover the integral peak load is not recommended.

To reduce costs, the HP is selected to cover only 70–75% of the maximum heat demand of the building. The rest of the heat demand is produced by an auxiliary

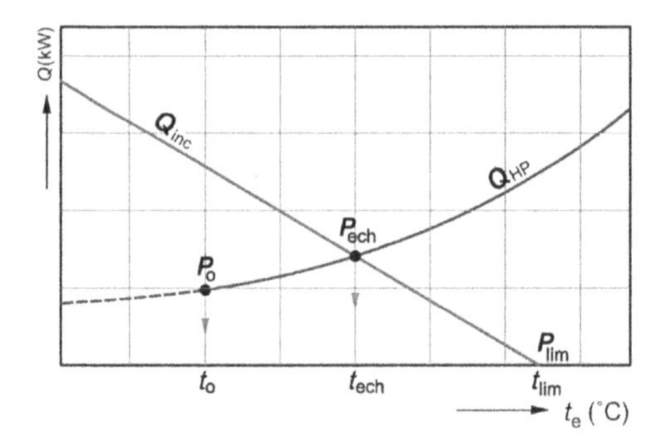

Figure 4.5 Heat demand provided in function of source temperature. P_o, *stop heating point;* P_{ech}, *balance point;* P_{lim}, *limit heating point.*

traditional source (i.e. electric heater or oil/gas boiler). In this case, the HP operates in bivalent mode (Figure 4.5), distinguishing three situations:

1. if the outdoor air temperature t_e is lower than the limit heating temperature t_{lim}, the HP provides the full heat demand up to the balance point temperature t_{ech}
2. if the outdoor air temperature t_e is lower than t_{ech}, the HP provides part of the heat demand and the rest is provided by the peak traditional source
3. when the outdoor air temperature t_e reaches the stop-heating temperature t_o, the HP is switched off and the traditional source meets the full heat demand.

Usually, the balance point corresponds to an outdoor temperature of 0 to +5°C. For temperate climate zones, the heat pump covers approximately two-thirds of the annual heat demand.

4.2.4 Water-to-Water Heat Pumps

Water-to-water HPs operate as GWHPs or SWHPs. Water-to-water HPs use water (e.g. aquifer-fed boreholes, lakes or water bodies) as the heat source and sink for heating and cooling. Antifreeze-to-water HPs are used in closed-loop ground-coupled installations. Water-to-water and antifreeze-to-water HPs are used for monovalent heating operation, as well as cooling, heat recovery and DHW production.

Heating/cooling changeover can be performed in the refrigerant circuit, but it is often more convenient to perform the switch in the water circuits. Several water-to-water HPs can be grouped together to create a central cooling and heating plant to serve several air-handling units. This application has advantages of better control, centralised maintenance, redundancy and flexibility.

4.2.5 Ground-Coupled Heat Pumps

Ground-coupled HPs use the ground as a heat source and sink. An HP may have a refrigerant-to-water heat exchanger or may be direct-expansion (DX). In systems with refrigerant-to-water heat exchangers, a water or antifreeze solution is pumped through horizontal, vertical or coiled pipes embedded in the ground. DX ground-coupled HPs use refrigerant in DX, flooded or recirculation evaporator circuits for the ground pipe coils. A loop of suitable pipe containing the refrigerant and lubricant is put in direct contact with the ground or water body. The compressor operation circulates the refrigerant directly around this loop, thus eliminating the heat transfer losses associated with the intermediate water-DX heat exchanger found in conventional water source HPs [4]. There is also no need for a source-side circulation pump as the compressor fulfils this role. However, care must be taken to ensure that the DX loops are totally sealed and corrosion resistant and that the lubricant is adequately circulated to meet the needs of the compressor.

A hybrid ground-coupled HP is a variation that uses a cooling tower or air-cooled condenser to reduce the total annual heat rejection to the ground coupling.

4.2.6 Hybrid Air-to-Water Systems

A hybrid air-to-water system integrates an air-to-water HP unit with another non-renewable heat source, such as a condensation gas boiler, to create a highly energy efficient heating and DHW system. This system can produce water flow temperatures from 25 up to 80 °C, making it suitable for any type of heat emitter, including radiant floor heating and radiators.

The intelligent hybrid HP measures the outdoor temperature, automatically adjusting the flow temperature to the emitters and calculating the efficiency of the HP. The system continuously evaluates whether the efficiency of the HP is higher than that of the condensing gas boiler. Based upon this evaluation, the energy source is selected, ensuring that the most efficient heat source is being used at all times. There are three operating conditions for this system [5]:

- HP only: for approximately 60% of the year, when the outdoor temperature is mild, the HP will supply energy for space heating. The primary energy-based efficiency in this mode is approximately 1.5.
- hybrid operation: for approximately 20% of the year, when outdoor temperatures are between −2 and 3 °C, the HP and the condensing gas boiler work together to provide energy for space heating. The system efficiency is approximately 1.0 in this mode.
- boiler only: when outdoor temperatures are below −2 °C, the condensation gas boiler provides the energy for space heating.

Throughout the year, the overall weighted primary energy efficiency is between 1.2 and 1.5, which is 30–60% higher compared with the best gas condensation boiler.

The hybrid HP system consists of three main components [5]:

1. The outdoor unit transmits the renewable energy extracted from the air to the indoor unit (hydro-box). The compact and whisper-quiet outdoor unit contains the inverter-driven compressor, which has a modulation ratio from approximately 20–100%. In partial load conditions, the outdoor heat exchanger is oversized, which increases the efficiency by up to 30%.
2. The hydro-box is mounted on the wall behind the condensing boiler. It contains the water-side elements of the system, such as the expansion tank and pump, as well as the controls for the system and the heat exchanger, which converts the renewable energy extracted from the air into hot water.
3. The condensing gas boiler is installed in front of the hydro-box. The combined dimensions of the boiler and hydro-box are approximately the same as a conventional wall-hung boiler.

The hybrid HP has been field tested in various climates and house types (i.e. size, age and energy rating) with a range of different heat emitters. The seasonal performance factor (SPF) measured during the winter of 2011–2012 varied between 1.25 and 1.6.

4.2.7 High-Temperature Heat Pumps

Most of the air-to-water HPs for residential applications use a single refrigerant cycle, which makes them a perfect solution for low water temperature applications in new buildings. Conversely, refurbishment requires higher water temperatures to work with existing high-temperature radiators. A new HP system has been developed [6] as an efficient high-water-temperature HP, which uses two refrigerant cycles with two different refrigerants (R410A and R134a) in a cascade system.

A single-stage HP is optimised to work with water temperatures of approximately 35 °C, ideal for radiant floor heating applications. The properties of the R410A refrigerant (see Chapter 3) match the requirements at these conditions very well. A low-temperature single-stage HP system is only able to produce maximum water temperatures of approximately 55 °C. Some medium-temperature single-stage HPs exist on the market, but use vapour injection resulting in two compression stages to reach higher operation pressures. Medium-temperature HPs working with a dual-stage cycle are able to reach water temperatures of approximately 65 °C. They can function with most of the existing radiators, which means that the investment cost and the potential damage to the building will be limited.

However, the advantages of the medium-temperature HPs are still limited compared with a condensing boiler. Medium-temperature HPs operating at a water temperature of 65 °C are pushed to the limit, resulting in rather low efficiencies at these operating conditions. This limitation is why a cascade high-temperature vapour compression-based HP was designed [6]. This type of HP operates on the refrigerants R410A and R134a. The combination of both refrigerants operating at their optimal conditions results in an HP system with high efficiency up to water temperatures of 80 °C over a large range of outdoor air temperatures, and down to −25 °C. The most important property of these HPs is their high efficiency at a leaving water temperature of 65 °C to operate in combination with existing radiators.

Figure 4.6 shows the schematic of a cascade HP system. The first refrigerant cycle with R410A runs from the outdoor unit to the indoor unit, and the second refrigerant cycle with R134a is located in the indoor unit. Both compressors in the outdoor and indoor units are inverter driven. The evaporator in the outdoor unit extracts heat from the outdoor air and exchanges this heat with the refrigerant cycle. After the evaporation of the refrigerant, the first inverter compressor elevates the pressure and temperature, delivering an intermediate condensation temperature. The heat of the R410A refrigerant cycle is exchanged via the first plate heat exchanger located in the indoor unit to the R134a refrigerant cycle. The second inverter compressor then delivers the second stage condensation temperature. Finally, R134a is condensed in the second plate heat exchanger, heating up the water circuit to a maximum temperature of 80 °C.

During operation, the intermediate condensation temperature of the R410A is kept at a semi-fixed value, allowing the R134a circuit to work independently from the outdoor air temperature. Balanced operation of the R134a circuit is possible because of

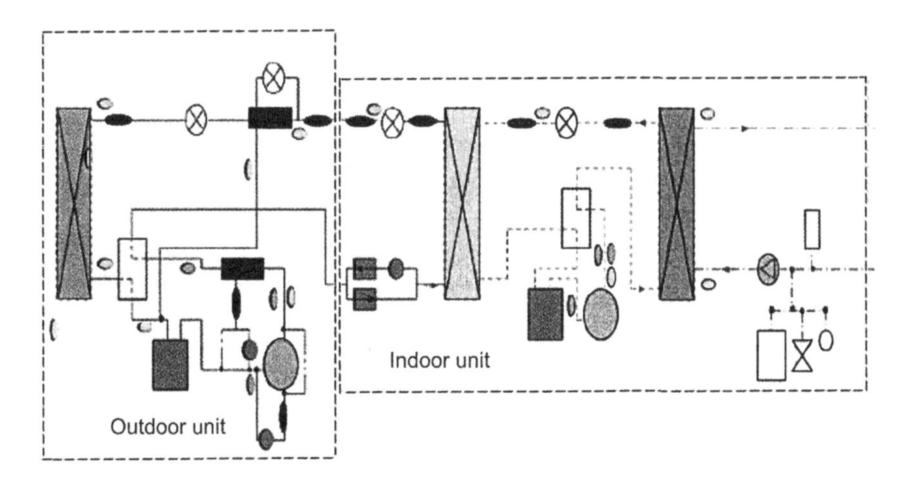

Figure 4.6 Schematic of a cascade heat pump.

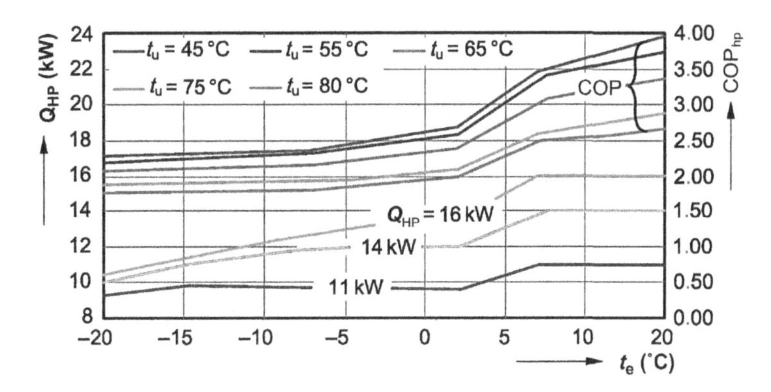

Figure 4.7 Thermal power and COP in function of outdoor air temperature.

the R410A controlled intermediate condensation temperature. The appropriate operation of the unit is focused on maintaining this intermediate condensation temperature.

The first dataset in Figure 4.7 shows the integrated thermal power Q_{HP} for three different capacity classes of HPs operating with a water temperature of 65 °C for outdoor air temperatures t_e, ranging from 20 to −20 °C. The second dataset shows the COP of the 14 kW HP model for water temperatures ranging from 45 up to 80 °C and outdoor air temperatures from 20 to −20 °C.

The DHW temperatures of up to 75 °C can be reached with the cascade system and without the assistance of an electrical heater. Figure 4.8 shows a simulation for the heating of DHW from 15 to 70 °C (measured at different levels in the tank) with an outdoor air temperature of 7 °C. The simulation shows an average COP of approximately 2.7. Even medium-temperature HPs have a large drop in efficiency when producing DHW close to 60 °C.

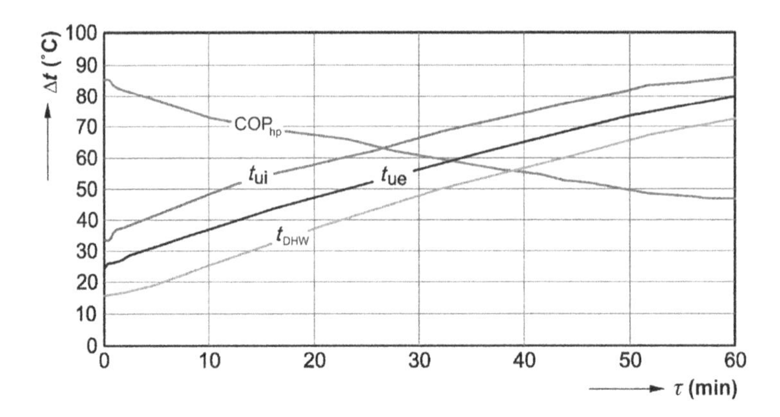

Figure 4.8 Simulation of efficiency for domestic hot-water heating.

Heating systems in older buildings operate at high water temperatures to be used in combination with high-temperature radiators. The new cascade HP system using two refrigerant cycles is now able to replace traditional boilers in existing buildings as they have the possibility to achieve high efficiencies for high water temperatures.

4.3 SELECTION OF HEAT SOURCE AND HEAT PUMP SYSTEM

An HP heating system consists of a heat source, an HP unit and a heat delivery system. For system planning, all of the components must be designed to interact optimally to ensure the highest level of performance and reliable operation [4].

As a rule, the heat source with the highest temperature levels should be selected to ensure the highest possible coefficient of performance and thereby the lowest operation costs.

If ground-water is available at a reasonable depth, temperature, acceptable quality and in sufficient quantity, the highest COP can be achieved. The best ground-water heat source system is an open system, which may require approval.

If the use of ground-water is not available, the ground can function as an efficient, effective thermal storage medium with a relatively high temperature. If sufficient surface area is available, horizontal collectors offer the most cost effective solution. If space is limited, vertical loops using geothermal energy can be effective. These heat sources are closed systems, meaning that the antifreeze solution (brine) stays within the buried tube system.

In direct expansion systems, the heat stored in the ground is absorbed directly by the refrigerant. Horizontal collectors are mainly used with this system.

If ground-water or ground source systems cannot be used, air as a heat source is available practically anywhere. These systems are particularly suitable for retrofits or combined with another heat source (i.e. bivalent operation).

The selection of the heat source determines to a great extent the required HP type and operation. The exact size selection of the HPs is important because oversized

Table 4.1 Specific Heat Demand for Building Heating

No.	Type of Building	Specific Heat Demand (W/m^2)
1	Passive house	15
2	Low energy building	40
3	New building with good insulation	50
4	Old building with standard insulation	80–90
5	Old building without insulation	120

systems operate with lower efficiencies, leading to excessive costs. The required thermal power of an HP must cover the total heat demand:

$$Q_{inc} = Q_h + Q_v \tag{4.18}$$

where Q_h is the heating demand and Q_v is the ventilation demand.

The determination of the heat demand can be performed according to European standard EN 12831 or national standards, for example DIN 4701 and EnEV 2002 (Germany); ÖNORM 7500 (Austria); SIA 380-1 and SIA 384.2 (Switzerland); and SR 1907 (Romania).

The values of the specific heat demand provided in Table 4.1 are based upon experience in European buildings. The specific heat demand is multiplied by the area of the heated space to give the total heat demand.

With active cooling, the refrigeration cycle of the HP is reversed using a four-way valve. The condenser is transformed into an evaporator and actively removes the unwanted heat (cooling) from the floor/wall areas and sends it into the ground.

The cooling capacity during the cooling operation is often higher than the heating requirement during heating operation. The determination of the cooling demand can be performed according to European standard EN 15243 or national standards; for example VDI 2078 (Germany) and SR 6648-1,2 (Romania).

Utilities sometimes offer a reduced price for electricity for an HP. In return, they maintain the right to interrupt the supply at certain periods during the day. The power supply can be interrupted three times for 2 h within a 24-h period. Therefore, the HP must meet the required daily heat demand in the time in which electricity is delivered. The required heat capacity of the HP Q_{HP}, in W, can be calculated as:

$$Q_{HP} = \frac{24Q_{inc}}{18 + 2} \tag{4.19}$$

To achieve maximum system efficiency, a separate, independent HP should be planned for DHW heating. This should be optimally sized and installed, and can provide additional functions (ventilation, cooling and dehumidification).

The heating HP is oversized for DHW heating during the summer and must operate at a higher temperature difference in the winter due to the water heating priority. The average DHW demand can be assumed to be 30 l/person/day at 45 °C. The maximum operating temperature in the distribution network is 60 °C.

If the heating system is to be used for DHW production, notice should be taken of the following [4]:

- A heating capacity of 0.25 kW per person for hot-water heating should be accounted for when sizing the HP (for a single family house).
- A hot-water tank for a three to five person household is approximately 300 l.
- Plan the heat exchanger for a 5-K temperature difference with a supply temperature of 55 or 65 °C. This results in an approximately 50 or 60 °C hot-water temperature produced by the HP.
- For internal heat exchangers, a surface area of 0.4–0.7-m^2/kW heating capacity is recommended. If this cannot be achieved, then plate heat exchangers will be required.

An exact calculation of the heat demand for DHW can be made using international or national norms.

Numerous compact units which combine heating, hot-water production and even controlled ventilation are on the market. In some cases, combination storage tanks are used in these applications. In these tanks, the domestic water and heating loop water are stored separately. The heat sources range from outdoor air to exhaust air to ground or ground-water. In many cases, various heat sources are combined.

With the use of a conventional combination storage tank, it should be noted that it is less efficient (and more costly) to heat the DHW via the heating water loop. This is because the HP is forced to operate with a higher temperature difference.

Once the total required heating capacity of the HP is determined, an appropriate HP model can be selected, based upon the technical characteristics data (see Chapter 5) provided by HP manufacturers.

4.4 DHW PRODUCTION FOR NEARLY ZERO-ENERGY BUILDINGS

In many cases, the HP is the preferred choice for a nearly zero-energy building (nZEB). Without entering into the various types of HPs that can be adopted (i.e. air-to-air, water-to-water, ground-source) as well as boiler-HPs, gas absorption and gas engine or entering into the calculations and feasibility studies necessary for each case, the main advantages of HPs are the possibility of producing DHW from heat recovery, the integration with solar thermal systems and good performance at part loads. The main problem to be faced when installing an HP for space heating into a nZEB is the production of the DHW [7].

Due to the many peculiarities of the DHW production within an nZEB context, such as the high thermal levels, the high design heating load (e.g. 18 kW for 10 l/min with a temperature difference of 15–40 °C), the problems related to the proliferation of legionella, and integration with renewable sources, if electric HPs are used, instantaneous production is not feasible and a storage volume is necessary to level the loads. The storage volume also allows adequate temperatures for both the use and the source (HP) sides. Basically, three plant schemes can be designed:

- a DHW storage with an internal coil (water heater): the storage volume is designed based on the DHW requirement, and the coil should be appropriately sized and the problem of legionella should be addressed

- a DHW storage with an external heat exchanger: the storage volume is designed based on the DHW requirement; the external heat exchanger allows higher efficiency while there are two pumps and the problem of legionella should still be addressed
- water storage and an instantaneous heat exchanger for DHW production: in this case, the storage volume is larger, there are always two pumps, but there is no risk of legionella proliferation.

4.5 INSTALLATION INSTRUCTIONS FOR HEAT PUMPS

In radiator heating systems, there are usually small amounts of water in the system. In such systems, a buffer tank should be added to avoid frequent switching (on and off) of the HP.

HPs can switch off during peak periods. Therefore, in a heating system with low thermal inertia, the buffer tank volume should be dimensioned so that the accumulated heat is sufficient to avoid breaks and cooling in the building. In systems with large volume, the buffer tank can be omitted, for example, in an indirect cooling system with a monovalent operating mode combined with radiant floor heating. These heaters must be installed with a differential pressure valve in the heating circuit manifold, which is located at the greatest distance from the HP to ensure a minimum amount of water pumped into closed heating circuits.

For the air-to-water HPs, the buffer tank is advantageous because, at a higher heat source temperature, the HP capacity increases and the heat demand decreases. The buffer tank role is to ensure operational periods long enough to prevent cyclic operation.

The buffer tank serves to separate hydraulic volumetric flow rates in the HP circuit and the heating circuit. If, for example, the heating circuit volume flow is reduced by thermostatic valves, then the volumetric flow rate of the HP circuit remains constant.

To ensure the minimum amount of heat carrier, HP systems without buffer tanks should be fitted with mixing valves. The heating circuit pump does not require variable speed but should have speed levels.

4.6 EXAMPLES OF HEAT PUMP UTILISATION

HPs are used for DHW and building heating (e.g. houses, office buildings, administrative buildings, hotels, hospitals, etc.), in industrial processes and agriculture. HPs can be used in both local and centralised heat supply systems.

The connection of the HP to the heating systems shows some features determined by their very different structure and heat consumption characteristics.

Figure 4.9 shows the schematic of an air-to-water HP used for heating and DHW production for a building with a floor surface of 170 m^2 and a heat demand of 16 kW. The heating is performed with fan coil units, and the water supply and return temperatures are 50 and 40 °C, respectively.

The installation is provided with heat storage and HP (I), which operates for heating until the outdoor temperature is −3 °C, taking heat from the outdoor air. Below this outdoor temperature, HP (II) begins to operate and takes the heat from the heat storage, which contains water heated by liquid refrigerant sub-cooling from the HP circuit.

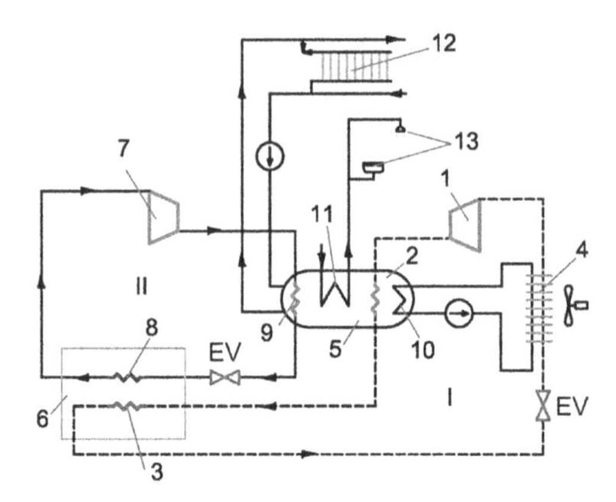

Figure 4.9 Air-to-water heat pump for a house heating and domestic hot-water (DHW) production. 1-Main compressor; 2-Condenser; 3-Subcooler; 4-Fan-cooling coil; 5-Fast heat storage; 6-Slow heat storage; 7-Second compressor; 8-Evaporator; 9-Condenser; 10-Defrosting cooling coil water heater; 11-DHW heat exchanger; 12-Fan coil units; 13-Shower and bathtub.

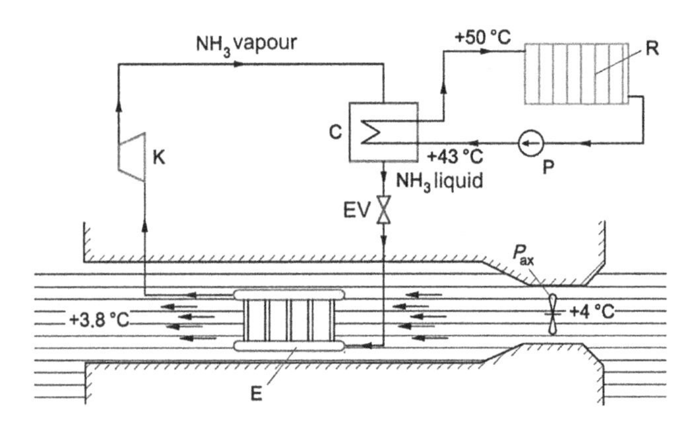

Figure 4.10 Water-to-water heat pump for administrative building.

The vaporisation temperature is −10 °C, and the condensation temperature is 55 °C. The working fluid used is R-22, and the absorbed electric powers by the two compressors are $P_I = 3.75$ kW and $P_{II} = 3.0$ kW.

One of the first uses of water-to-water HPs was in Switzerland for the heating of the City Hall building in Zurich [8]. The functional scheme of the system is shown in Figure 4.10. The working fluid of the HP is ammonia (NH_3), and the heat source is river water with a temperature in winter of approximately 4 °C. The heat taken up in the evaporator E is yielded in the condenser C to the water radiator circuit R to a temperature of approximately 50 °C.

For an outdoor air temperature of −20 °C and a room temperature of 18 °C, the heat demand is 175 kW. The HP only covers the basic heat load of 80 kW for offices and accommodation heating. The remaining heat demand is covered by an electric heater with a power of 65 kW. The plant is designed to operate with the HP for

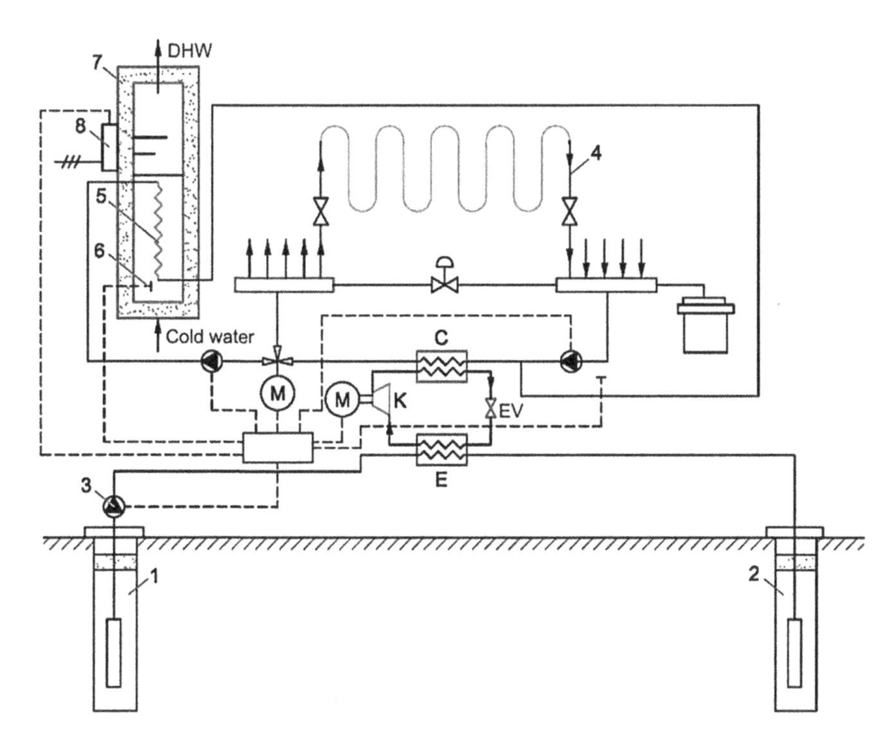

Figure 4.11 Water-to-water heat pump for house and domestic hot-water (DHW) heating using geothermal water heat source. 1-Geothermal production well; 2-Injection well; 3-Geothermal water circulation pump; 4-Radiant floor heating; 5-DHW coil; 6-Hot-water thermostat; 7-Hot-water storage; 8-Additional electrical resistance.

approximately 1500 h/year. During the summer, the system can be used to cool the air from the ventilation circuit of meeting rooms.

In the local heat supply, building systems commonly use the geothermal water source HPs for low water temperatures (those below 40 °C).

Water-to-water HPs with the scheme shown in Figure 4.11 are used to heat multi-family buildings and to produce DHW. They are used with radiant floor heating systems (4). The production of the domestic hot-water is made with an immersed coil (5) in the storage tank (7), which is additionally equipped with an electrical resistance heater (8) for periods of maximum hot-water consumption. In the summer season when heating is unnecessary, the HP operates exclusively for DHW production.

To maintain the productivity parameters of the geothermal source and to maintain ecological balance, the geothermal water must be returned, after cooling in the HP evaporator, to an injection well.

Figure 4.12 presents the scheme of an HP system used for heating an individual home. The heat source is the ground, from which heat is absorbed by vertical loops that circulate an antifreeze solution (water-glycol). The antifreeze solution flowing through the HP evaporator is sent to vertical ground heat exchangers (loops). The installation includes a storage tank of approximately 10 m^3, where heat may be stored for the periods in which the heating system is not running. A system of two three-way valves allows for heating directly from the HP condenser or from the storage tank or both simultaneously.

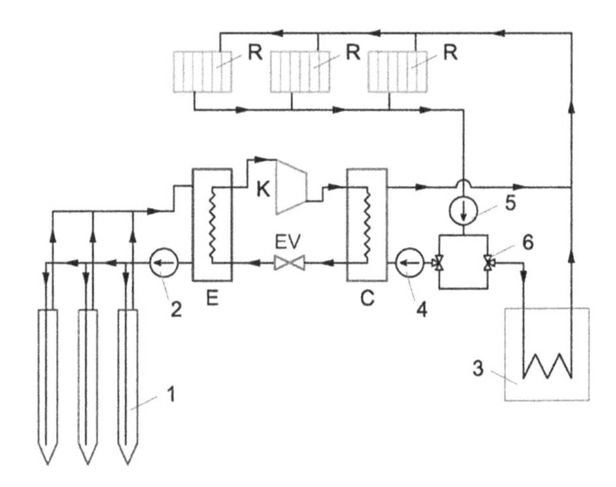

Figure 4.12 Ground-coupled heat pump for a house heating. 1-Ground loops; 2-Circulation pump; 3-Storage tank; 4-Hot-water pump; 5-Heating system pump; 6-Three-way valve.

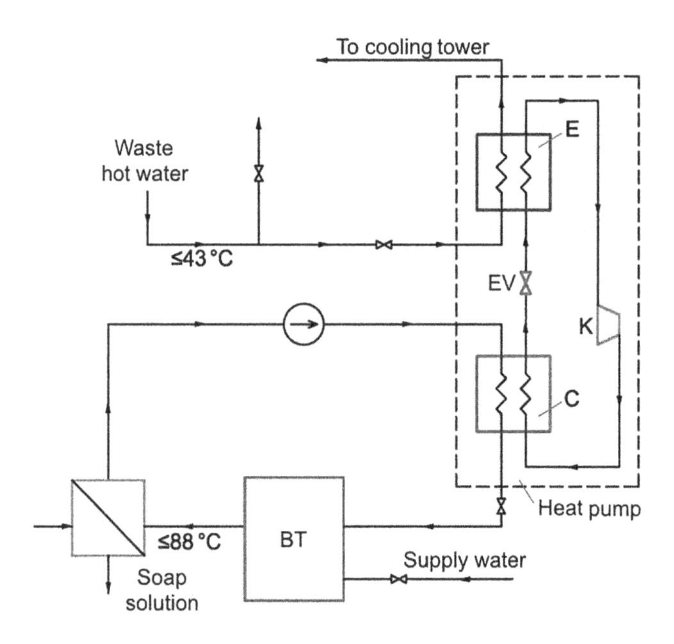

Figure 4.13 Air-to-air heat pump used in a copper pipe factory.

In all of the HP systems, those with applications in industry are more numerous, characterised by higher power corresponding to the waste energy discharged by industrial technological systems.

Figure 4.13 illustrates the functional scheme of a water-to-water HP, used in a factory that produces copper pipes.

The water resulting from cooling the extrusion mould has a temperature of approximately 43 °C. Instead of cooling the water in a cooling tower for reuse, this water is employed as a heat source for the evaporator E of an HP. The condenser C produces hot water with a temperature of approximately 88 °C, which is accumulated in a buffer tank (BT) of approximately 45,000 l. This hot-water is used to heat the soap solution used for the copper pipe lubrication during the manufacturing process. The thermal power of this HP is 1600 kW, the power absorbed by the compressor is 460 kW, and the COP is 3.4.

In the case of vapour compression-based refrigeration systems, the refrigerant is situated at the end of compression, in a superheated vapour state, with a temperature higher than the ambient air. One possibility to value the condensation heat and sensitive heat of superheated vapours represents their use by HP systems.

Figure 4.14 shows a coupling between an HP and a refrigeration system for a milk factory. The system consists of two cascade refrigeration circuits. The cold water production stage includes the compressor K_1 and the evaporator E_1, which operates with R-22. The condensation of R-22 vapour occurs in the pipes of the intermediate heat exchanger C_1–E_2, which also has part of the evaporator role in the hot-water production stage. The second stage operates with R-114 and also has a compressor K_2 and a condenser C_2. The condensate sub-cooler SC recovers the sub-cooling heat to produce DHW at 60 °C.

For an annual system operation of 6000 h/year, the fuel savings achieved is 36 toe.

Currently, a large number of installations are based on the use of solar energy in combination with the HP. For the most efficient use of solar energy throughout the year, the energy is stored in summer and is consumed during the winter. The heat from the sun is stored in water contained in thermal-insulated storage tanks.

The heating and hot-water system in a solar house (Figure 4.15) from Essen, Germany uses solar energy as a heat source through the collector plane (5).

The hot-water from the solar collector circuit heats boiler (1) used for DHW production and boiler (2) used to heat the water for the radiators. When this system is not

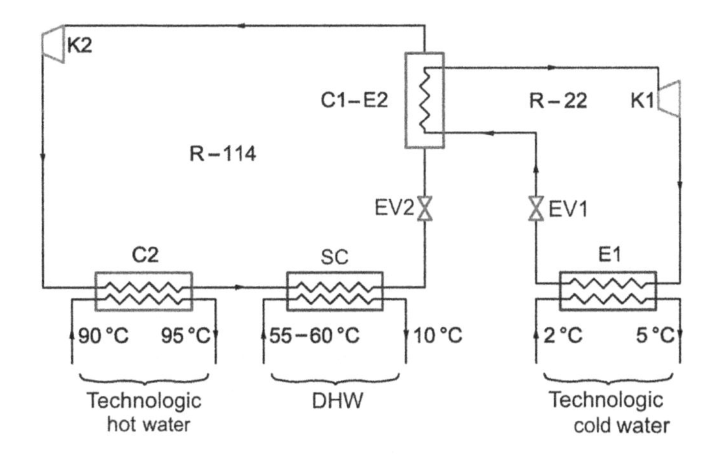

Figure 4.14 Water-to-water heat pump coupled with a refrigeration system for technological heating/cooling.

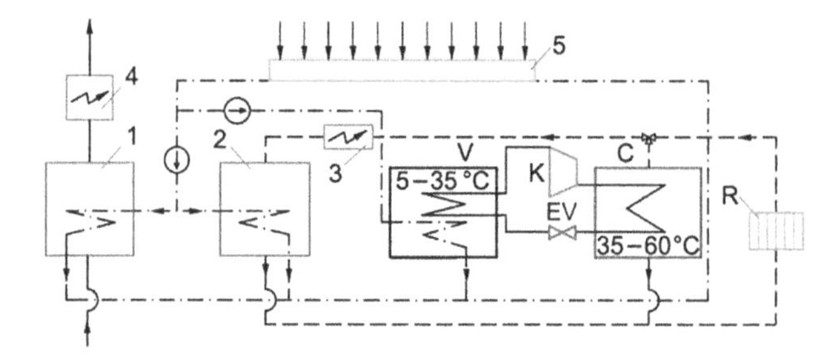

Figure 4.15 Solar-assisted water-to-water heat pump for heating and domestic hot-water (DHW). 1-DHW storage; 2-Heating water storage; 3-Additional electric heater; 4-DHW electric heater; 5-Flat solar collector.

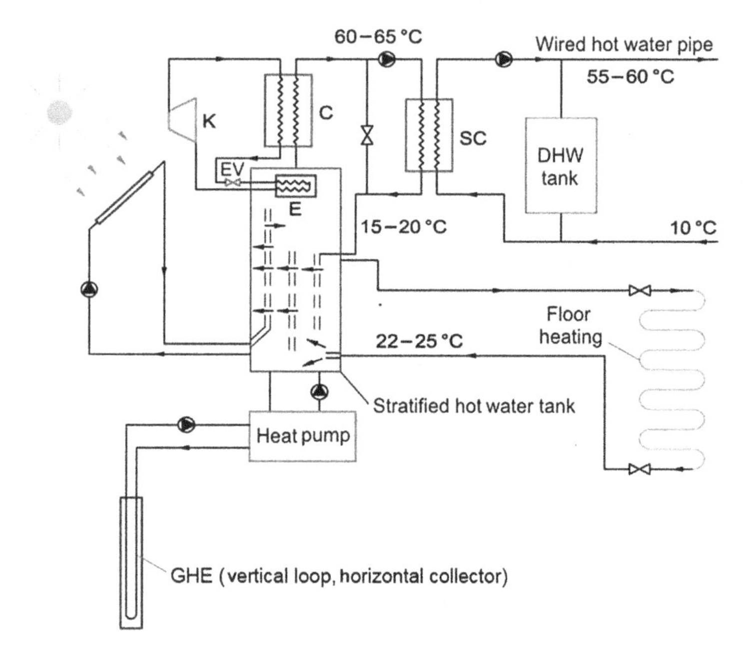

Figure 4.16 Radiant floor heating system with heat pump and solar collector.

enough, an HP starts to operate using heated water from the solar collector, producing the hot water in condenser C for the heaters. The installation has the possibility to heat water with electric energy which is not possible with solar energy.

Figure 4.16 presents a radiant floor heating system with an HP and a vacuum tube solar collector, operating with water temperatures of 20–30 °C. A ground-water HP with horizontal collectors or vertical loops and solar collectors transfers its heat to a stratified hot-water tank. Reheating hot-water is performed both by an additional HP and directly by an electric heater.

For a family house with a heat demand of approximately 8000 kWh/year, the solar collector can cover approximately 2000 kWh/year. If taking into account the circulation pumps' power, the electricity consumption of the HP reaches 1500 kW/year, and the COP of the HP can reach a value of 4.

Initially, installation can be realised without the solar collector, leaving the possibility of installing one in the future. Instead, an HP for hot water reheating can be mounted with an electric heater.

4.7 RENEWABLE ENERGY SOURCE CONTRIBUTION FROM HP SALES IN EU

Member States are obliged to set trajectories on how to achieve their mandatory renewable energy source (RES) targets for 2020, which is performed via their National Renewable Energy Action Plan (NREAP). The assessment of all plans shows a target contribution from HPs towards the 2020 total energy use of 1298 TWh [9]. Ambition among Member States is not evenly spread, however. While the UK aims to cover 36% of its RES target by contributions from HPs, other countries such as Portugal, Bulgaria, Estonia, Malta and Romania have not included the technology in their plans [10].

Today's contribution of the HP stock can be determined from industry sales statistics. A total of 4.5 million HPs were sold from 2005 to 2011 (Figure 4.17). The actual stock is higher, as significant HP markets existed before 2005 in Austria, France, Germany, Italy, Sweden and Switzerland. Using average values for SPF and E_U, the total annual contribution of the HP stock is 27.37 TWh, saving 6.83 Mt of GHG emissions annually [11].

Additionally, the growing share of intermittent renewable energy sources from solar and wind requires peak shaving and storage options for better integration. HPs are well equipped to provide demand-side potential, thus indirectly helping to further increase the share of renewable sources in the energy mix.

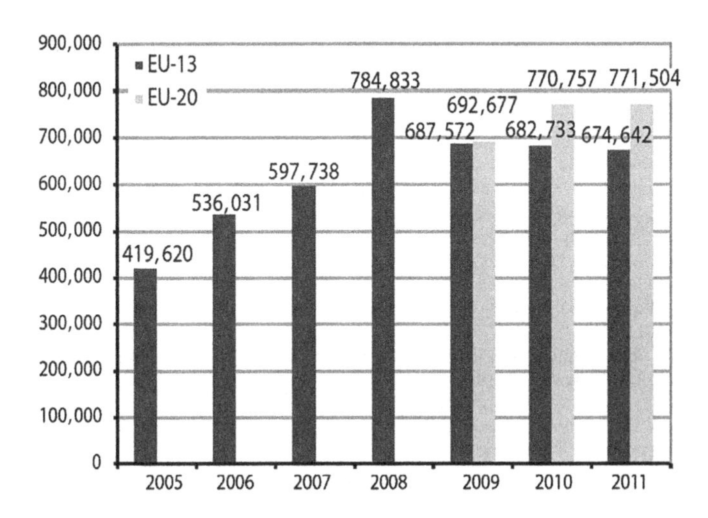

Figure 4.17 Development of heat pump sales from 2005 to 2011.

4.8 CONCLUSIONS

Correct adaptation of the heat source and the heating system for the operating mode of HPs leads to safe and economical operation of the heating system.

Heating systems with HPs consume minimum energy during operation and are certainly a solution for energy optimisation of buildings.

The HP mode requires some additional investments. If the capacity of the HP is selected as larger than the condensing capacity in the pure refrigeration mode, the additional capacity costs must be covered by the savings in energy costs.

An HP provides the necessary technical conditions for efficient use of solar heat for heating and DHW production.

A combined cooling and heating system with an HP is always more effective than a traditional system if its requirements are taken into the consideration in the design process. For renovation, the applicability is more limited and is always case dependent.

The main barrier for the use of HPs for retrofitting is the high distribution temperature of conventional heating systems in existing residential buildings. Traditional design temperatures of up to 70–90 °C are too high for the present HP generation, with a maximum economically acceptable heat distribution temperature of approximately 55 °C. In addition to the application of existing HPs in already improved standard buildings with reduced heat demand, the development and market introduction of new high-temperature HPs is a major task for the replacement of conventional heating systems with HPs in existing buildings.

REFERENCES

[1] Bloch HP, Hoefner JJ. Reciprocating compressors: operation and maintenance. Houston, TX: Gulf Publishing Company; 1996.
[2] Japikse D, Baines NC. Introduction to turbomachinery. Oxford: Oxford University Press; 1977.
[3] Bush JW, Elson JP. Scroll compressor design criteria for residential air conditioning and heat pump applications. In: Proceedings of the 1988 international compressor engineering conference, July 1988, vol. 1. p. 83–92.
[4] Ochsner K. Geothermal heat pumps: a guide to planning & installing. London-Sterling: Earthscan; 2007.
[5] Aspeslagh B, Debaets S. Hybrid heat pumps – saving energy and reduction carbon emissions. Rehva J 2013;50(2):20–5.
[6] Steen D, Logghe J. New technology for high temperature heat pumps. Rehva J 2009;46(5):31–6.
[7] Fabrizio E. System for zero energy houses. Rehva J 2014;51(6):7–10.
[8] Radcenco V, Florescu Al, Duica T, Burchiu N, Dimitriu S, et al. Heat pumps systems. Bucharest: Technical Publishing House; 1985 [in Romanian]
[9] Nowak T. Heat pumps – a renewable energy technology. Rehva J. 2011;48(4):10–12.
[10] EREC: Mapping renewable energy pathways towards 2020. <http://www.repap2020.eu/>, Brussels; 2011.
[11] Nowak T. Heat pumps in Europe – a 'smart' future? In: Proceedings of IEA heat pump conference, <http://www.hpc2011web.org/>; 2011.

CHAPTER 5

Ground-Source Heat Pump Systems

5.1 GENERALITIES

Recently, the ground-source heat pump (GSHP) system has attracted more and more attention due to its superiority of high energy efficiency and environmental friendliness [1–4]. Renewable forms of energy such as solar, wind, biomass, hydro and earth energy produce low or zero greenhouse gas (GHG) emissions. The temperature of the ground is fairly constant below the frost line. The ground is warmer in the middle of winter and cooler in the middle of summer than the outdoor air. Thus, the ground is an efficient heat source. A GSHP system includes three principle components: (i) a ground connection subsystem, (ii) heat pump (HP) subsystem, and (iii) heat distribution subsystem.

The GSHPs comprise a wide variety of systems that may use groundwater, ground, or surface water as heat sources or sinks. These systems have been basically grouped into three categories by ASHRAE [5]: (i) groundwater heat pump (GWHP) systems, (ii) surface water heat pump (SWHP) systems, and (iii) ground-coupled heat pump (GCHP) systems. The schematics of these different systems are shown in Figure 5.1. Many parallel terms exist: e.g. geothermal heat pump, earth energy system, and ground-source system.

The GWHP system, which utilises groundwater as heat source or sink, has some marked advantages including a low initial cost and minimal requirement for ground surface area over other GSHP systems [6]. However, a number of factors seriously restrict the wide application of the GWHP systems, such as the limited availability of groundwater and the high maintenance cost due to fouling corrosion in pipes and equipment. In a SWHP system, heat rejection/extraction is accomplished by circulating working fluid through high-density polyethylene (HDPE) pipes positioned at an adequate depth within a lake, pond, reservoir or the suitable open channels. The major disadvantage of the system is that the surface water temperature is more affected by weather conditions, especially in winter.

Among the various GSHP systems, the vertical GCHP system has attracted the greatest interest in research fields and practical engineering. Several literature reviews on the GCHP technology have been reported [7,8].

In a GCHP system, heat is extracted from or rejected to the ground via a closed-loop, i.e. ground heat exchanger (GHE), through which pure water or antifreeze fluid circulates. The GHEs commonly used in the GCHP systems typically consist of HDPE pipes which are installed in either vertical boreholes (called vertical GHE) or horizontal trenches (horizontal GHE). In direct expansion systems, the heat stored in the ground is absorbed directly by the working fluid (refrigerant). This results in an increased coefficient of performance (COP). Horizontal GHEs are mainly used for this system.

Ground-Source Heat Pumps. DOI: http://dx.doi.org/10.1016/B978-0-12-804220-5.00005-9

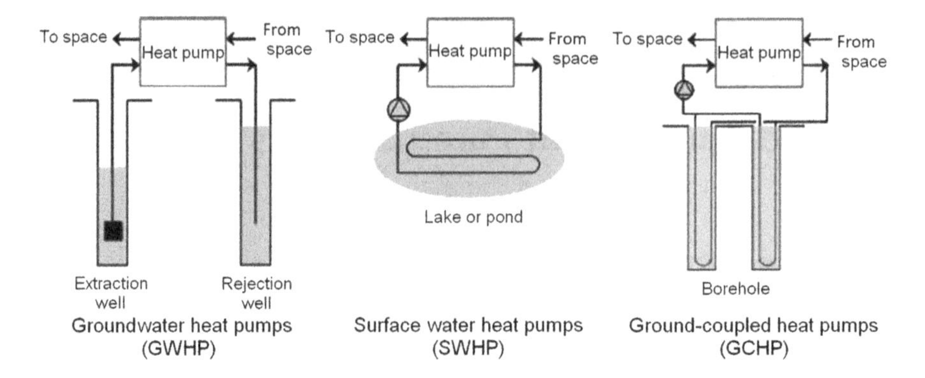

Figure 5.1 Schematic of different ground-source heat pumps.

The GSHPs work best with heating systems, which are optimised to operate at lower water temperature than radiator and radiant panel systems (floor, wall and ceiling). GSHPs have the potential to reduce cooling energy by 30–50% and reduce heating energy by 20–40% [9]. The GSHPs tend to be more cost-effective than conventional systems in the following applications:

- in new constructions where the technology is relatively easy to incorporate, or to replace an existing system at the end of its useful life
- in climates characterised by high daily temperature swings, or where winters are cold or summers hot, and where electricity cost is higher than average
- in areas where natural gas is unavailable or where the cost is higher than electricity.

This chapter presents a detailed description of GSHPs and their development, discusses the most common simulation models and programs of vertical GHEs currently available, and describes different applications of the models and programs. Additionally, a new GWHP using a heat exchanger with special construction, tested in a laboratory, with the possibility of obtaining better energy efficiency with combined heating and cooling by GCHP is presented. Finally, the advanced engineering applications of hybrid GCHP systems are also briefly analysed.

5.2 DESCRIPTION OF SWHP SYSTEMS

Surface water bodies can be very good heat sources and sinks, if properly used. The maximum density of water occurs at 4.0 °C, not at the freezing point of 0 °C. This phenomenon, in combination with the normal modes of heat transfer to and from takes, produces a temperature profile advantageous to efficient HP operation. In some cases, lakes can be the very best water supply for cooling. Various water circulation systems are possible and several of the more common are presented next.

In a closed-loop system, a water-to-air HP is linked to a submerged coil. Heat is exchanged to or from the lake by the refrigerant circulating inside the coil. The HP transfers heat to or from the air in the building.

In an open-loop system, water is pumped from the lake through a heat exchanger and returned to the lake some distance from the point at which it was removed.

The pump can be located either slightly above or submerged below the lake water level. For HP operation in the heating mode, this type is restricted to warmer climates. Entering lake water temperature must remain above 5.5 °C to prevent freezing.

Thermal stratification of water often keeps large quantities of cold water undisturbed near the bottom of deep lakes. This water is cold enough to adequately cool buildings by simply being circulated through heat exchangers. An HP is not needed for cooling, and energy use is substantially reduced. Closed-loop coils may also be used in colder lakes. Heating can be provided by a separate source or with HPs in the heating mode. Pre-cooling or supplemental total cooling are also permitted when the water temperature is between 10 and 15 °C.

Advantages of closed-loop SWHPs are (i) relatively low cost because of reduced excavation costs, (ii) low pumping energy requirements and (iii) low operating cost. Disadvantages are (i) the possibility of coil damage in public lakes and (ii) wide variation in water temperature with outdoor conditions.

5.3 DESCRIPTION OF GWHP SYSTEMS

A GWHP system removes groundwater from a well and delivers it to an HP (or an intermediate heat exchanger) to serve as a heat source or sink [5]. Both unitary and central plant designs are used. In the unitary type, a large number of small water-to-air HPs are distributed throughout the building. The central plant uses one or a small number of large-capacity chillers supplying hot and chilled water to a two- or four-pipe distribution system. The unitary approach is more common and tends to be more energy efficient.

Direct systems (in which groundwater is pumped directly to the HP without an intermediate heat exchanger) are not recommended except on the very smallest installations. Although some installations of this system have been successful, others have had serious difficulty even with groundwater of apparently benign chemistry. The specific components for handling groundwater are similar. The primary items include (i) wells (supply and, if required, injection), (ii) a well pump (usually submerged), and (iii) a groundwater heat exchanger. The use of a submerged pump avoids the possibility of introducing air or oxygen into the system. A back-washable filter should also be installed. The injection well should be located from 10 to 15 m in the downstream direction of the groundwater flow.

In an open-loop system, the intermediate heat exchanger between the refrigerant and the groundwater is subject to fouling, corrosion and blockage. The required flow rate through the intermediate heat exchanger is typically between 0.027 and 0.054 l/s. The groundwater must either be reinjected into the ground by separate wells or discharged to a surface system such as a river or lake. To avoid damage due to corrosion, the conductivity of the water should not exceed 450 micro Siemens per cm.

Commissioning must be performed by the manufacturer's customer service department or qualified representatives. The heat collection, heat distribution and electrical connections must all be in working order before commissioning.

The drill diameter should be at least 220 mm (and larger for sandy conditions to prevent sand entry).

The groundwater flow rate G must be capable of delivering the full capacity required from the heat source. This depends on the evaporator cooling power Q_0 and the water cooling degree and is given by the following equation:

$$G = \frac{Q_0}{\rho_w c_w (t_{wi} - t_{we})} \tag{5.1}$$

in which ρ_w is the water density; c_w is the specific heat of water; and t_{wi} and t_{we} are the water temperatures at the HP inlet and the HP outlet, respectively.

The values of the flow rate for each HP model are usually provided in the manufacturer's data sheets.

The technical characteristics of water-to-water HPs produced by the German company Stulz GmbH are presented in Table 5.1 [10] as an example.

Table 5.2 summarises the calculated COP values of GWHP and SWHP systems, operating as water-to-water HPs.

The 'Geotherm' system [4] uses a specially built heat exchanger (Figure 5.2) placed in an extraction well with a 1.0 m diameter and a depth of 2.0 m. The heat exchanger is mounted between a GCHP and a groundwater source with a reduced flow rate and of any water quality. This heat exchanger consists of a set of four coaxial coils made of HDPE tubes with a diameter of 25 mm, immersed in a cylindrical reservoir made of

Table 5.1 Technical Characteristics of Stulz Water-to-Water Heat Pumps

Characteristics	MU	Type of Heat Pump						
		WWP71	WWP121	WWP181	WWP221	WWP271	WWP442	WWP912
Thermal power	kW	6.6	12.2	18.8	22.1	27.1	43.9	90.8
Electrical power	kW	1.25	2.2	3.35	3.8	4.9	8.0	17.1
COP	–	5.2	5.6	5.6	5.7	5.5	5.5	5.3
Length	mm	700	700	700	700	700	700	1300
Breadth	mm	550	550	550	550	850	850	850
Height	mm	700	700	700	700	700	700	850
Acoustic level	dB(A)	32	32	32	32	38	36	46
Refrigerant	–	R-407C						

Table 5.2 The COP of Water-to-Water GWHP and SWHP Systems

Water Temperature at Evaporator Inlet t_s (°C)	Water Temperature at Condenser Outlet, t_u (°C)				
	30	35	40	45	50
5	4.55	4.10	3.70	3.40	3.15
10	5.30	4.65	4.15	3.75	3.45
15	6.25	5.35	4.70	4.20	3.85
20	7.70	6.35	5.45	4.80	4.30
25	9.95	7.80	6.45	5.55	4.85
30	14.10	10.10	7.95	6.55	5.60

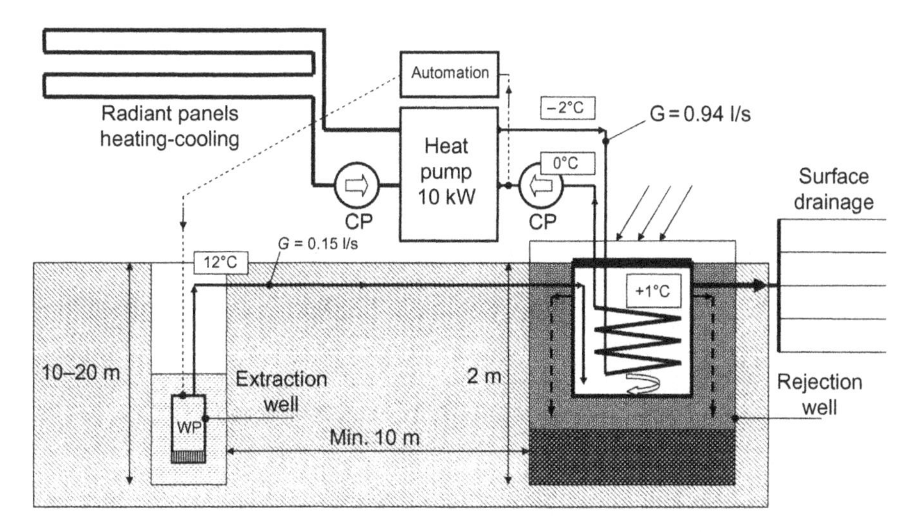

Figure 5.2 Schematic of 'Geotherm' GWHP system.

glass fibre reinforced resins (0.8 m diameter and 1.2 m height) supplied with groundwater at the bottom side.

The HP used in conjunction with the intermediate heat exchanger is a GCHP system of 10 kW with a COP of 4.

The secondary circuit of the heat exchanger (towards the HP) circulates an antifreeze solution (glycol 20%), which enters the HP at 0 °C and leaves at −2 °C, transported by a circulation pump with a flow rate of 0.94 l/s. The glycol flow in the tubes is ensured by the circulation pump within the HP. Outside of the coils, the groundwater from the cylindrical reservoir is involved in a flow among the spires of coils by a submersible pump. The relatively small pressure loss on the secondary circuit of the heat exchanger allows the use of a reduced power circulation pump for the glycol.

In the primary circuit of the heat exchanger (outside the tubes), groundwater enters with a temperature of 12 °C and is evacuated to approximately 1 °C (in heating mode). Because the temperature drop is 11 °C, compared with 4 °C in the usual systems, it is possible to obtain the same thermal power with a groundwater flow rate nearly three times lower. The pressure loss on the primary circuit of the heat exchanger is 26 kPa for the mentioned flow rate. The heat exchange is realised mainly by the groundwater supply, and the heat exchanged directly with the ground around the extraction well is also important. The heat transfer surface is 20 m^2, and the heat transfer coefficient is 154 W/m^2 K.

The groundwater is then evacuated through the top of the heat exchanger by the gravity in the rejection well. If the rejection well cannot retrieve all of the groundwater flow rate, surface drainage through a network of perforated pipes buried at 50–80 cm or another evacuation solution (lake, river or sewer) is recommended.

Regardless of the outdoor air and ground temperature, the HP will always operate at the same optimum temperatures because of the automation.

The automation starts the groundwater inlet (electro-valve or submersible pump) only when the return water-glycol temperature goes below 1 °C. The groundwater flow rate is limited to 4–12 l/min depending on the thermal power of the HP (4–12 kW).

During the summer, the intermediate heat exchanger can operate in a passive cooling mode in which the HP only produces domestic hot water (DHW) using heat recovered from the air-conditioned (A/C) space. In this case, the heat carrier from the heaters is transported with the circulation pump directly to the intermediate heat exchanger.

The main advantages of this Geotherm HP system with intermediate heat exchanger are the following:

- lower HP installation costs
- very deep wells or long trenches on large surfaces are not necessary
- eliminated drilling of extraction and rejection wells
- reduced groundwater flow rate (30–40% of the usual flow rate)
- the location of the Geotherm heat exchanger in the absorption well improves the thermal state through direct heat exchange with the ground; and
- the groundwater quality is not important because heat transfer is made by HDPE.

The installation of a GWHP that uses a safety refrigerant is possible in any space that is both dry and protected from freezing temperatures. The system should be installed on an even, flat surface and the construction of a free-standing base is recommended. The placement of the unit should be such that servicing and maintenance are possible. Generally, only flexible connections to the HP should be implemented.

5.4 DESCRIPTION OF GCHP SYSTEMS

The ground serves as an ideal heat source for monovalent HP systems. The GCHP is a subset of the GSHP and is often called a closed-loop HP. A GCHP system consists of a reversible vapour-compression cycle that is linked to a GHE buried in the soil (Figure 5.1). The heat transfer medium, an antifreeze solution (brine), is circulated through the GHE (collector or loop) and the HP by an antifreeze solution pump. The GHE size needs to take into account the total annual heating demand, which, for domestic heating operation, is typically between 1700 and 2300 h in central Europe.

5.4.1 Ground Characteristics and Conditions

The most important thermal ground characteristics are thermal conductivity (λ), density (ρ), specific heat (c) and thermal volumetric capacity ($C = \rho \times c$). Table 5.3 summarises the values of thermal conductivity and thermal capacity for different ground types. The accumulation capacity and thermal conductivity of the ground are functions of the ground moisture content and the quantity of minerals.

If groundwater flows through the area, heat exchange is more complex because the water flow determines the changes of the temperature field in the direction of the water movement, enhancing heat exchange. The hydraulic conductivity of the ground is thus

Table 5.3 Thermal Conductivity and Volumetric Thermal Capacity of Different Grounds

Ground Type	Thermal Conductivity λ (W/(m K))	Thermal Capacity C (MJ/(m^3 K))
Igneous rocks		
Basalt	1.3–2.3	2.3–2.6
Diorite	2.0–2.9	2.9
Gabbros	1.7–2.5	2.6
Granite	2.1–4.1	2.1–3.0
Peridotite	3.8–5.3	2–7
Rhyolite	3.1–3.4	2.1
Metamorphic rocks		
Gneiss	1.9–4.0	1.8–2.4
Marble	1.3–3.1	2.0
Meta quartzite	5.8	2.1
Mica schist	1.5–3.1	2.2
Clay schist's	1.5–2.1	2.2–2.5
Sedimentary rocks		
Chalk	2.5–4.0	2.1–2.4
Marl	1.5–3.5	2.2–2.3
Quartzite	3.6–6.6	2.1–2.2
Salt	5.3–6.4	1.2
Hone	1.3–5.1	1.6–2.8
Argillaceous rock	1.1–3.5	2.1–2.4
Unconsolidated soils		
Dry gravel	0.4–0.5	1.4–1.6
Saturated gravel	1.8	2.4
Moraine	1.0–2.5	1.5–2.5
Dry sand	0.3–0.8	1.3–1.6
Saturated sand	1.7–5.0	2.2–2.9
Dry clay	0.4–1.0	1.5–1.6
Saturated clay	0.9–2.3	1.6–3.4
Peat	0.2–0.7	0.5–3.8
Other substances		
Bentonite	0.5–0.8	3.9
Cement	0.9–2.0	1.8
Ice (−10 °C)	2.32	1.87
Plastic	0.39	–
Dry air (10–20 °C)	0.02	0.0012
Steel	60	3.12
Water (+10 °C)	0.58	4.19

another parameter to consider. This requires determining the groundwater velocity and the well-known horizontal pressure gradient:

$$v = ki \tag{5.2}$$

where v is the Darcy velocity, in m/s; i is the pressure gradient; and k is the hydraulic conductivity, in m/s.

Table 5.4 presents the influence of hydraulic conductivity on the ground thermal characteristics. The GCHP is further subdivided according to GHE type: horizontal GHE and vertical GHE.

The ambient climatic conditions affect the temperature profile below the ground surface and need to be considered when designing a heat exchanger. The ground temperature distribution is affected by the structure and physical properties of the ground, the ground surface cover (e.g. bare ground, lawn, snow, etc.) and the climate interaction (i.e. boundary conditions) determined by air temperature, wind, solar radiation, air humidity and rainfall. The temperature distribution at any depth below the earth's surface remains unchanged throughout the year, and the temperature increases with depth with an average gradient of approximately 30 °C/km.

Heat flux, which is a gauge of the amount of thermal energy coming out of the earth, is calculated by multiplying the geothermal gradient by the thermal conductivity of the ground. Each rock type has a different thermal conductivity, which is a measure of the ability of a material to conduct heat. Rocks rich in quartz, such as sandstone, have a high thermal conductivity, indicating that heat readily passes through them. Rocks that are rich in clay or organic material, such as shale and coal, have low thermal conductivity, meaning that heat passes more slowly through these layers. If the heat flux is constant throughout a drill hole (i.e. water is not flowing up or down the hole), then it is obvious that low-conductivity shale layers will have a higher geothermal gradient compared to high-conductivity sandstone layers.

From the point of view of the temperature distribution, Popiel et al. [11] distinguish three ground zones:

1. Surface zone reaching a depth of approximately 1 m, in which the ground temperature is very sensitive to short time changes of weather conditions.
2. Shallow zone extending from the depth of approximately 1–8 m (for dry, light soils) or 20 m (for moist, heavy, sandy soils), where the ground temperature is almost constant and close to the average annual air temperature; in this zone, the ground temperature distributions depend mainly on the seasonal cycle weather conditions.
3. Deep zone (below approximately 8–20 m), where the ground temperature is practically constant (and rises very slowly with depth according to the geothermal gradient).

Table 5.4 Influence of the Hydraulic Conductivity on Thermal Parameters

Ground Type	k (m/s)	λ (W/(m K))		C (MJ/(m^3 K))	
		Dry	Saturated	Dry	Saturated
Clay	10^{-8}–10^{-10}	0.2–0.3	1.1–1.6	0.3–0.6	2.1–3.2
Silt	10^{-5}–10^{-8}	0.2–0.3	1.2–2.5	0.6–1.0	2.1–2.4
Sand	10^{-3}–10^{-4}	0.3–0.4	1.7–3.2	1.0–1.3	2.2–2.4
Gravel	10^{-1}–10^{-3}	0.3–0.4	1.8–3.3	1.2–1.6	2.2–2.4

For positioning the horizontal collector or a vertical loop, it is important that the ground is well settled and level. The more water the ground contains, the better the heat transmission. A smaller ground area is required for dense, wet soil than for dry, crumbling ground. Rainwater is very important for the regeneration of the ground. The ground must settle before heat is collected. Limited ice build-up around the collector tubing is allowable and even preferred (for better heat transfer).

5.4.2 Types of Horizontal GHEs

Horizontal GHEs (Figure 5.3) can be divided into at least three subgroups: single-pipe, multiple-pipe and spiral. Single-pipe horizontal GHEs consist of a series of parallel pipe arrangements laid out in trenches. Typical installation depths in Europe vary from 0.8 to 1.5 m. Consideration should be given to the local frost depth and the extent of snow cover in winter. Horizontal GHEs are usually the most cost-effective when adequate yard space is available and the trenches are easy to dig. Antifreeze fluid runs through the pipes in a closed system. The values of the specific extraction/rejection power q_E for ground [12,13] are given in Table 5.5. For a specific power of extraction/rejection q_E, the required ground area A can be obtained as follows [14]:

$$A = \frac{Q_0}{q_E} \tag{5.3}$$

where $Q_0 = Q_{HP} - P_e$ is the cooling power of HP.

The values of the cooling power for each HP model are usually provided in the manufacturer's data sheets.

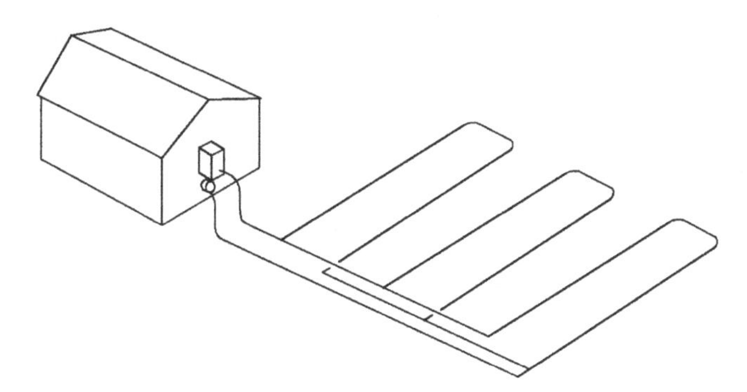

Figure 5.3 Horizontal ground heat exchanger.

Table 5.5 Specific Extraction Power for Ground

No.	Type of Ground	q_E (W/m^2)
1	Dry sandy	10–15
2	Moist sandy	15–20
3	Dry clay	20–25
4	Moist clay	25–30
5	Ground with groundwater	30–35

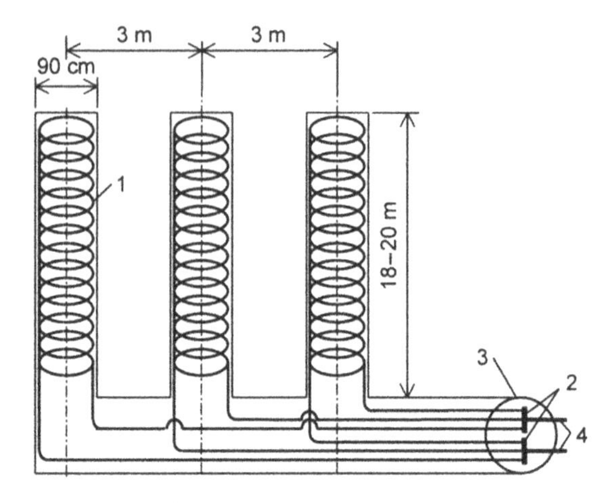

Figure 5.4 Spiral ground coil.

It is important to ensure that the pipe is not crimped or squeezed. The collector pipes should be laid on and covered with a protective layer of sand before covering them with ground.

To save required ground area, some special GHEs have been developed [15]. Multiple pipes (two, four or six), placed in a single trench, can reduce the amount of required ground area. The trench collector is vilely used in North America, and less in Europe.

The spiral loop (Figure 5.4) is reported to further reduce the required ground area. This consists of pipe unrolled in circular loops in trenches with a horizontal configuration. For the horizontal spiral loop layout, the trenches are generally a depth of 0.9–1.8 m. The distance between coil tubes is of 0.6 – 1.2 m. The length of collector pipe is of 125 m per loop (up to 200 m). The ends of parallel coils 1 are arranged by a manifold-collector 2 in a heart 3, and then the antifreeze fluid is transported by main pipes 4 at HP. For the trench collector, a number of pipes with small diameters are attached to the steeply inclined walls of a trench several metres deep.

Horizontal ground loops are the easiest to install while a building is under construction. However, new types of digging equipment allow horizontal boring, thus making it possible to retrofit such systems into existing houses with minimal disturbance of the topsoil and even allowing loops to be installed under existing buildings or driveways.

For all horizontal GHEs in heating-only mode, the main thermal recharge is provided by the solar radiation falling on the earth's surface. Therefore, it is important not to cover the surface above the ground heat collector.

Disadvantages of the horizontal systems are: (i) these systems are more affected by ambient air temperature fluctuations because of their proximity to the ground surface, and (ii) the installation of the horizontal systems needs much more ground area than vertical systems.

5.4.3 Types of Vertical GHEs

There are two basic types of vertical GHEs or borehole heat exchangers (BHEs): U-tube and concentric- (coaxial-) tube system configurations (Figure 5.5).

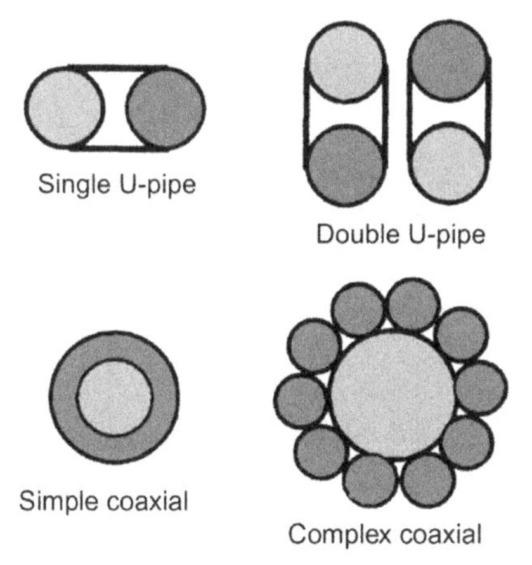

Figure 5.5 Common vertical GHE designs.

BHEs are widely used when there is a need to install sufficient heat exchanger capacity under a confined surface area, such as when the earth is rocky close to the surface, or where minimum disruption of the landscape is desired.

The U-tube vertical GHE may include one, tens, or even hundreds of boreholes, each containing single or double U-tubes through which heat exchange fluids are circulated. Typical U-tubes have a nominal diameter in the range of 20–40 mm and each borehole is normally 20–200 m deep, with a diameter ranging from 100 to 200 mm. Concentric pipes, either in a very simple method with two straight pipes of different diameters or in complex configurations, are commonly used in Europe. The borehole annulus is generally backfilled with some special material (grout) that can prevent contamination of groundwater.

Geological analysis should be completed before drilling to give an indication of the underground layers and an exact collection capacity. The drilling and the insertion of the loop pipe should be completed by a specialised and licensed drilling company. The hole will be refilled and the tubing secured according to industry standard and should include proper sealing in the case of groundwater.

A typical borehole with a single U-tube is illustrated in Figure 5.6. The required borehole length L can be calculated by the following steady-state heat transfer equation [5]:

$$L = \frac{qR_g}{t_g - t_f} \tag{5.4}$$

where: q is the heat transfer rate, in kW; t_g is the ground temperature, in K; t_f is the heat carrier fluid (i.e. antifreeze, refrigerant) temperature, in K; R_g is the effective thermal resistance of ground per unit length, in (m K)/kW.

The GHE usually are designed for the worst conditions by considering that the need to handle three consecutive thermal pulses of various magnitude and duration: yearly average ground load q_a for 20 years, the highest monthly ground load q_m for

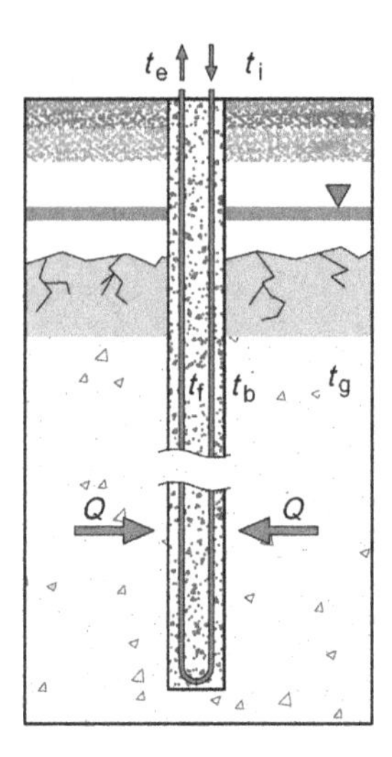

Figure 5.6 Schematic of a vertical grouted borehole.

1 month, and the peak hourly load q_h for 6 h. The required borehole length to exchange heat at these conditions is given by [16]:

$$L = \frac{q_h R_b + q_a R_{20a} + q_m R_{1m} + q_h R_{6h}}{t_g - (t_f + \Delta t_g)} \tag{5.5}$$

where: R_b is the effective borehole thermal resistance; R_{20a}, R_{1m}, R_{6h} are the effective ground thermal resistances for 20-year, 1-month, and 6-h thermal pulses; Δt_g is the increase of temperature due to long-term interference effect between the borehole and adjacent boreholes.

The effective ground thermal resistance depends mainly on the ground thermal conductivity and, to a lesser extent, on the borehole diameter and the ground thermal diffusivity. Alternative methods of computing the thermal borehole resistance are presented by Bernier [16] and Hellström [17].

The antifreeze solution (brine) circulation loop normally consists of the following components: collector pipes, manifold, vent, circulation pump, expansion vessel, safety valve, insulation or condensate drainage and flexible connections to the HP unit (closed system).

All of the components should be corrosion-free materials and should be insulated with closed cell insulation in the heating space to prevent condensation.

The piping should be sized such that the fluid velocity does not exceed 0.8 m/s.

The antifreeze mass flow rate must be capable of transporting the full thermal capacity required from the heat source. The mass flow rate m_b, in kg/s, is given by:

$$m_b = \frac{3600 Q_0}{c_b \Delta t} \tag{5.6}$$

where Q_0 is the cooling power of the HP, in kW; c_b is the specific heat of antifreeze, in kJ/kg K; and Δt is the temperature difference, in K (e.g. 3 K).

The ground loop circulation pump must be sized to achieve the minimum flow rate through the HP.

The nominal volume V_N of the diaphragm expansion vessel, in litres, for the antifreeze solution circulation loop can be calculated as:

$$V_N = \frac{p_{si} + 0.05}{p_{si} - (p_{st} + 0.05)}(\beta + 0.005)\, V_T \tag{5.7}$$

where p_{si} is the safety valve purge pressure, equal to 0.3 MPa; p_{st} is the nitrogen preliminary pressure (0.05 MPa); β is the thermal expansion coefficient ($\beta = 0.01$ for Tyfocor); and V_T is the total volume of the system (heat exchanger, inlet duct, HP), in litres.

As an example, the theoretical characteristics of ground-to-water HPs produced by the international companies Stulz GmbH from Germany and Ochsner GmbH from Austria are summarised in Tables 5.6 and 5.7 [10,18].

The installation of a GCHP that uses a safety refrigerant is possible in any space that is both dry and protected from freezing temperatures. The system should be

Table 5.6 Technical Characteristics of Stulz Ground-to-Water Heat Pumps

Characteristics	MU	Type of Heat Pump						
		SWP71	SWP91	SWP111	SWP171	SWP2118	SWP332	SWP702
Thermal power	kW	6.8	9.1	11.4	16.5	20.9	32.4	68.0
Electrical power	kW	1.6	2.1	2.6	3.6	4.7	7.7	16.2
COP	–	4.3	4.4	4.4	4.6	4.4	4.2	4.2
Length	mm	700	700	700	700	700	700	1300
Breadth	mm	550	550	550	550	850	850	850
Height	mm	700	700	700	700	700	700	850
Acoustic level	dB(A)	32	32	32	32	38	36	46
Refrigerant	–	R-407C						

Table 5.7 Technical Characteristics of Ochsner Ground-to-Water Heat Pumps

Characteristics	MU	Type of Heat Pump						
		GMSW9	GMSW11	GMSW15	GMSW21	GMSW28	GMSW33	GMSW38
Thermal power	kW	6.5	8.2	11.0	15.0	19.5	24.2	28.3
Electrical power	kW	1.45	1.85	2.45	3.4	4.4	5.5	6.4
COP	–	4.4	4.4	4.5	4.4	4.4	4.4	4.4
Length	mm	1150	1150	1150	1150	1150	1150	1150
Breadth	mm	400	400	600	600	600	600	600
Height	mm	650	650	650	650	650	650	650
Refrigerant	–	R-407C						

installed on an even, horizontal surface and the construction of a free-standing base is recommended. The placement of the unit should be such that servicing and maintenance are possible. Generally, only flexible connections to the HP should be implemented.

Advantages of the vertical GCHP are that it (i) requires relatively small ground area, (ii) is in contact with soil that varies very little in temperature and thermal properties, (iii) requires the smallest amount of pipe and pumping energy, and (iv) can yield the most efficient GCHP system performance. The disadvantage is a higher cost because of the expensive equipment needed to drill the borehole.

5.4.4 Simulation Models of GHEs

The main objective of the GHE thermal analysis is to determine the temperature of the heat carried fluid, which is circulated in the U-tube and the HP, under certain operating conditions. Actually, the heat transfer process in a GHE involves a number of uncertain factors, such as the ground thermal properties, the groundwater flow rate and building loads over a long lifespan of several or even tens of years. In this case, the heat transfer process is rather complicated and must be treated, on the whole, as a transient one. In view of the complication of this problem and its long-time scale, the heat transfer process may usually be analysed in two separated regions. One is the solid soil/rock outside the borehole, where the heat conduction must be treated as a transient process. Another sector often segregated for analysis is the region inside the borehole, including the grout, the U-tube pipes and the circulating fluid inside the pipes. This region is sometime analysed as being steady-state and sometime analysed as being transient. The analyses on the two spatial regions are interlinked on the borehole wall. The heat transfer models for the two separate regions are as follows.

5.4.4.1 Heat Conduction Outside Borehole

A number of simulation models for the heat transfer outside the borehole have been recently reported, most of which were based on either analytical methodologies or numerical methods [19].

Kelvin's line-source. The earliest approach to calculating the thermal transport around a heat exchange pipe in the ground is Kelvin's line-source theory, i.e. the infinite line source [20]. According to Kelvin's line-source theory, the temperature response in the ground due to a constant heat rate is given by:

$$t(r,\tau) - t_0 = \frac{q}{4\pi\lambda}\int_{r^2/4a\tau}^{\infty} \frac{e^{-u}}{u}\,du \tag{5.8}$$

where: r is the distance from the line-source and τ the time since the start of the operation; t is the temperature of the ground at distance r and time τ; t_0 is the initial temperature of the ground; q is the heating rate per length of the line source; λ and a are the thermal conductivity and diffusivity of the ground.

The solution to the integral term in Eqn (5.8) can be found from the related references [21,22]. It was estimated that using Kelvin's line source could cause a noticeable error when $a\tau/r_b^2 < 20$.

Cylindrical source model. The cylindrical source solution for a constant heat transfer rate was developed by Carslaw and Jaeger [23], then refined by Ingersoll et al. [21], and later employed in a number of research studies [24,25]. In this model, the borehole is assumed as an infinite cylinder surrounded by homogeneous medium with constant properties (ground). It also assumes that the heat transfer between the borehole and ground with perfect contact is of pure heat conduction.

Based on the governing equation of the transient heat conduction along with the given boundary and initial conditions, the temperature distribution of the ground can be given in the cylindrical coordinate:

$$\begin{aligned} &\frac{\partial^2 t}{\partial r^2}+\frac{1}{r}\frac{\partial t}{\partial r}=\frac{1}{a}\frac{\partial t}{\partial \tau} \quad r_b<r<\infty \\ &-2\pi r_b\lambda\frac{\partial t}{\partial \tau}=q \quad r=r_b,\ \tau>0 \\ &t-t_0=0 \quad \tau=0,\ r>r_b \end{aligned} \tag{5.9}$$

where r_b is the borehole radius.

The cylindrical source solution is given as follows:

$$t-t_0=\frac{q}{\lambda}G(z,p) \tag{5.10}$$

where $z=a\tau/r_b$, $p=r/r_b$.

As defined by Carslaw and Jaeger [23], the expression $G(z, p)$ is only a function of time and distance from the borehole centre. The temperature on the borehole wall, where $r=r_b$, i.e. $p=1$, is of interest because it is the representative temperature in the design of GHEs. However, the expression $G(z, p)$ is relatively complex and involves integration from zero to infinity of a complicated function, which includes some Bessel functions. Fortunately, some graphical results and tabulated values for the $G(z, p)$ function at $p=1$ are available in some related references [21]. An approximate method for G was proposed by Hellström [17].

Eskilson's model. Both the one-dimensional model of Kelvin's theory and the cylindrical source model neglect the axial heat flow along the borehole depth. Eskilson [26] made major progress to account for the finite length of the borehole. In this model, the ground is assumed to be homogeneous with constant initial and boundary temperatures, and the thermal capacity of the borehole elements, such as the pipe wall and the grout, are neglected. The basic formulation of the ground temperature is governed by the heat conduction equation in cylindrical coordinates:

$$\begin{aligned} &\frac{\partial^2 t}{\partial r^2}+\frac{1}{r}\frac{\partial t}{\partial r}+\frac{\partial^2 t}{\partial z^2}=\frac{1}{a}\frac{\partial t}{\partial \tau} \\ &t(r,0,\tau)=t_0 \\ &t(r,z,0)=t_0 \\ &q(\tau)=\frac{1}{L}\int_D^{D+L} 2\pi r\lambda \left.\frac{\partial t}{\partial r}\right|_{r=r_b} dz \end{aligned} \tag{5.11}$$

in which L is the borehole length; D signifies the uppermost part of the borehole, which can be thermally neglected in engineering practice.

In Eskilson's model, the numerical finite-difference method is used on a radial-axial coordinate system to obtain the temperature distribution of a single borehole with finite length. The final expression of the temperature response at the borehole wall to a unit step heat pulse is a function of τ/τ_s and r_b/L only:

$$t_b - t_0 = -\frac{q}{2\pi\lambda} f(\tau/\tau_s, r_b/L) \tag{5.12}$$

where $\tau_s = L^2/9a$ means the steady-state time. The *f*-function is essentially the dimensionless temperature response at the borehole wall, which was computed numerically.

The disadvantage of this approach is its time-consuming nature, and it can hardly be incorporated directly into a design and energy analysis program for practical application, since the *f*-functions of the GHEs with different configurations have to be precomputed and stored in the program as a massive database.

Finite line-source solution. Based on the Eskilson's model, an analytical solution to the finite line source has been developed by a research group [27] that considers the influences of the finite length of the borehole and the ground surface as a boundary. Some necessary assumptions are taken in the analytical model in order to derive an analytical solution:

- the ground is regarded as a homogeneous, semi-infinite medium with constant thermo-physical properties
- the boundary of the medium (ground surface) keeps a constant temperature t_0, same as its initial one, throughout the time period concerned
- the radial dimension of the borehole is neglected so that it may be approximated as a line-source stretching from the boundary to a certain length L
- as a basic case of study, the heating rate per length of the source q is constant since the starting instant $\tau = 0$.

The solution of the temperature excess was given by Zeng et al. [27]:

$$t(r,z,\tau) - t_0 = \frac{q}{4\pi\lambda}\int_0^L \left[\frac{erfc\left(\frac{\sqrt{r^2+(z-l)^2}}{2\sqrt{a\tau}}\right)}{\sqrt{r^2+(z-l)^2}} - \frac{erfc\left(\frac{\sqrt{r^2+(z+l)^2}}{2\sqrt{a\tau}}\right)}{\sqrt{r^2+(z+l)^2}} \right] dl \tag{5.13}$$

Equation (5.13) demonstrates that the temperature on the borehole wall, where $r = r_b$, varies with time and borehole length. The temperature at the middle of the borehole length ($z = L/2$) is usually chosen as its representative temperature. An alternative is the integral mean temperature along the borehole length, which may be determined by numerical integration of Eqn (5.13). For the convenience of applications, the former is usually accepted as the representative temperature in the design and analysis program.

Other typical numerical models. Hellström [17] proposed a simulation model for ground heat stores, which are densely packed ground loop heat exchangers used for seasonal thermal energy storage. This type of system may be directly used to heat buildings with or without an HP. Muraya et al. [28] developed a transient finite-element model of the heat transfer around a vertical U-tube heat exchanger for a GCHP system to study the thermal interference that occurred between the adjacent legs of the U-tube. Rottmayer et al. [29] presented a finite difference model that simulated the heat transfer process of a U-tube heat exchanger. A geometric factor was introduced to account for the noncircular geometry used to represent the pipes in the borehole.

5.4.4.2 Heat Transfer Inside Borehole

The thermal resistance inside the borehole, which is primarily determined by thermal properties of the grouting materials and the arrangement of flow channels of the borehole, has a significant impact on the GHE performance. The main objective of this analysis is to determine the entering and leaving temperatures of the circulating fluid in the borehole according to the borehole wall temperature, its heat flow and the thermal resistance.

One-dimensional model. A simplified one-dimensional model has been recommended for GHE design, which considers the U-tube as a single 'equivalent' pipe [22,30]. In this model, both the thermal capacity of the borehole and the axial heat flux in the grout and pipe walls are negligible as the borehole dimensional scale is much smaller compared with the infinite ground outside the borehole. Thus, the heat transfer in this region is approximated as a steady-state, one-dimensional process.

Two-dimensional model. Hellström [17] derived the analytical two-dimensional solutions of the thermal resistances among pipes in the cross-section perpendicular to the borehole axis, which is superior to empirical expressions and the one-dimensional model. In the two-dimensional model, the temperature of the heat carrier fluid in the U-tubes is expressed as a superposition of the two separate temperature responses caused by the heat fluxes per unit length q_1 and q_2 from the two pipes of the U-tube, as shown in Figure 5.7. If the borehole wall t_b is considered as uniform along the borehole length, the fluid temperatures in the U-tubes can be obtained as follows:

$$\begin{cases} t_{f1} - t_b = R_{11}q_1 + R_{12}q_2 \\ t_{f2} - t_b = R_{12}q_1 + R_{22}q_2 \end{cases} \tag{5.14}$$

where R_{11} and R_{22} are the thermal resistances between the circulating fluid in each pipe and the borehole wall, and R_{12} is the resistance between the two pipes. A linear transformation of Eqn (5.14) leads to:

$$\begin{cases} q_1 = \dfrac{t_{f1} - t_b}{R_1^\Delta} + \dfrac{t_{f1} - t_{f2}}{R_{12}^\Delta} \\[2ex] q_2 = \dfrac{t_{f2} - t_b}{R_2^\Delta} + \dfrac{t_{f2} - t_{f1}}{R_{12}^\Delta} \end{cases} \tag{5.15}$$

where:

$$R_1^\Delta = \frac{R_{11}R_{22} - R_{12}^2}{R_{22} - R_{12}}, \quad R_2^\Delta = \frac{R_{11}R_{22} - R_{12}^2}{R_{11} - R_{12}}, \quad R_{12}^\Delta = \frac{R_{11}R_{22} - R_{12}^2}{R_{12}} \tag{5.16}$$

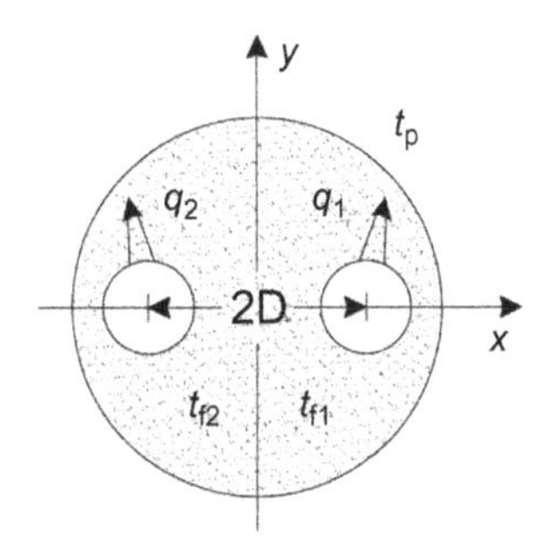

Figure 5.7 Configuration of a U-tube in a borehole.

For the instance of the symmetric disposal of the U-tube inside the borehole ($R_{11} = R_{22}$), these resistances can be deduced as:

$$R_1^{\Delta} = R_2^{\Delta} = R_{11} + R_{12} \text{ and } R_{12}^{\Delta} = (R_{11}^2 - R_{12}^2)/R_{12} \tag{5.17}$$

It is notable that there is no distinction between the entering and exiting pipes because this model does not take into account the heat transmission on the axial flow of the circulating fluid. In this case, Eskilson made the following assumptions to simplify the problem [13]: $t_{f1} = t_{f2} = t_f$ and $q_1 = q_2 = q/2$. Therefore, the thermal resistance between the fluid and borehole wall can be determined by:

$$R_{b2} = \frac{R_{11} - R_{12}}{2} \tag{5.18}$$

With the aid of these assumptions the temperatures of the fluid entering and exiting the GHE can be calculated. However, the temperatures of the fluid circulating through different legs of the U-tubes are, in fact, different. As a result, the thermal interference between the U-tube legs is inevitable, which degrades the effective heat transfer in the GHEs. With the assumption of an identical temperature for all the pipes, it is impossible for the two-dimensional model to reveal the impact of this thermal interference on the GHE performance.

Quasi-three-dimensional model. On the basis of the two-dimensional model previously mentioned, a quasi-three-dimensional model was proposed by Zeng et al. [31], which takes account of the fluid temperature variation along the borehole depth. To keep the model concise and analytically manageable, the conductive heat flux in the grout in the axial direction, however, is neglected. The energy balance equations can be written for up-flow and down-flow of the heat carrier fluid as:

$$\begin{cases} -m_f c_f \dfrac{dt_{f1}}{dz} = \dfrac{(t_{f1} - t_b)}{R_1^{\Delta}} + \dfrac{(t_{f1} - t_{f2})}{R_{12}^{\Delta}} \\ m_f c_f \dfrac{dt_{f2}}{dz} = \dfrac{(t_{f2} - t_b)}{R_2^{\Delta}} + \dfrac{(t_{f2} - t_{f1})}{R_{12}^{\Delta}} \end{cases} \quad (0 \leq z \leq L) \tag{5.19}$$

where m_f is the mass flow rate of fluid, and c_f is the specific heat of fluid.

Two conditions are necessary to complete the solution:

$$\begin{cases} z = 0, & t_{f1} = t_f' \\ z = L, & t_{f1} = t_{f2} \end{cases} \tag{5.20}$$

The general solution of this problem is derived by Laplace transformation, which is slightly complicated in form. For the instance of the symmetric placement of the U-tube inside the borehole, the temperature profiles in two pipes were illustrated by Diao et al. [32]. For the purpose of practical applications an alternative parameter $\varepsilon = (t_i - t_e)/(t_e + t_b)$ is derived from the temperature profiles, which is known as the heat transfer efficiency of the borehole. It should be noted that t_i and t_e are the entering/exiting fluid temperatures to/from the U-tube (Figure 5.6). A more accurate heat conduction resistance between the fluid inside the U-tube and the borehole wall can be calculated by:

$$R_{b3} = \frac{L}{m_f c_f}\left(\frac{1}{\varepsilon} - \frac{1}{2}\right) \tag{5.21}$$

Table 5.8 Comparison of the Current Models of GHEs

Specification	Model	Method	Interference Between Boreholes	Boundary Effects
Outside borehole	Kelvin's line source	Infinite line source	Yes	No
	Cylindrical source	Infinite cylindrical source	Yes	No
	Eskilson's model	Combination of numerical and analytical methods	Yes	Yes
	Finite line-source solution	Analytical method	Yes	Yes
Specification	Model	Method	Interference between U-tube pipes	Heat flux along depth
Inside borehole	One-dimensional	–	No	No
	Two-dimensional	–	Yes	No
	Quasi-three-dimensional	–	Yes	Yes

The quasi-three-dimensional model was validated and recommended for the design and thermal analysis of the GHEs.

5.4.4.3 Comparisons of the Analytical and Numerical Models

Although the numerical models can offer a high degree of flexibility and accuracy compared with the analytical models, most of the numerical models using polar or cylindrical grids may be computationally inefficient due to a large number of complex grids. In addition, it is difficult to directly incorporate the numerical models into a design and energy analysis program unless the simulated data are precomputed and stored in a massive database.

The analytical models are usually based on a number of assumptions and simplifications to solve the complicated mathematical algorithms; therefore, the accuracy of analytical results is slightly reduced due to the assumption of the line source at the centre of the borehole, which neglects the physical size of the U-tube in the borehole. However, the required computation time of the analytical model is much shorter than with the numerical models. Another advantage of the analytical models is that the straightforward algorithm deduced from the analytical models can be readily integrated into a design or simulation program. A summary of the characteristics of the numerical and analytical models of the GHEs [19] is given in Table 5.8.

5.4.5 Computer Programs for GCHP Design/Simulation

The reliability and stability of a GHE design mainly depends on its ability to reject or extract heat to/from ground over a long-term period and avoidance of excessive heat build up or heat loss in the ground. A good design program for the GCHPs should have high computational efficiency, which allows for the calculation of the transient effects over long timespans. Actually, there are numerous uncertain factors which to some extent, affect the final sizing of a GHE, such as the employed mathematical methodology, the allowed minimum/maximum temperatures of the fluid entering to the HP, the properties of the ground, borehole configuration and net annual energy transfer to the ground. However, the mathematical methodology or the heat transfer model of the GHEs is the crucial part for a design program. A number of design tools for vertical GHEs based on some typical heat transfer models have been developed in the past two decades.

5.4.5.1 Design Programs Based on the Line-Source Model

- *The Earth Energy Designer (EED) program.* The algorithms of the Lund programs [17,26] were developed based on Eskilson's approach where the temperature response of the borehole field is converted to a set of non-dimensional temperature response factors (*f*-functions). Those *f*-functions depend on the spacing between the boreholes on the ground surface and the borehole length. The *f*-function values obtained from the numerical simulations have been stored in a data file, which is accessed for rapid retrieval of data by the computer programs. A major drawback of the programs is that the users need to have a good knowledge of the input parameters and do some advance calculations.

  To make the Lund programs easier to use, the EED program has been developed on the same basis as the previous computer programs. The fluid temperature of the GHEs is calculated according to the monthly heating and cooling loads and the borehole thermal resistance. The thermal properties of the ground, the pipe materials and the heat carrier fluid are saved in a database. However, for cases with changing borehole length or borehole distance, the program needs to interpolate between suitable *f*-functions, and the interpolation process causes computing errors.
- *The GLHEPRO program* was developed for designing vertical GHEs used in commercial or institutional buildings [19]. Eskilson's approach is the basis for this program. The design methodology is based on a simulation that predicts the temperature response of the GHEs to monthly heating and cooling loads and monthly peak heating and cooling demands over a number of years. The temperature of the fluid inside the pipes in the borehole is determined using a one-dimensional, steady-state borehole thermal resistance. The design procedure involves the automatic adjustment of the GHE size to meet user-specified minimum or maximum HP entering fluid temperatures.
- *The GEOSTAR program* was developed for the design and simulation of GHEs by a research group in China [33]. This software is able to size GHEs to meet user-specified minimum and maximum HP entering fluid temperatures for a given set of design conditions, such as building load, ground thermal properties, borehole configuration and HP operating characteristics. The GHE heat transfer models employed consist of two regions: the heat conduction process of the solid soil or rock outside of the borehole and the region inside the borehole. For the first region, an explicit analytical solution of the finite line source in a semi-finite medium is derived for convenient calculation of the thermal resistance outside the borehole for long periods. Assuming the same heat transfer rate per unit length of each borehole, the borehole wall temperature of each individual borehole in a GHE can be obtained by means of an analytical solution. For the heat transfer insider the borehole, the quasi-three-dimensional model that takes into account the fluid temperature variation along the borehole length is used to calculate the fluid temperatures of the up-flow and down-flow channels. The analysis of the two

spatial regions is interlinked on the borehole wall. In addition, the modelling procedure uses spatial superimposition for multiple boreholes and sequential temporal superimposition to determine the arbitrary heating or cooling loads of the systems.

- *The EnergyPlus and eQUEST programs*, which are popular building energy simulation programs, were extended to include GCHP system simulations. Models of a water-source HP and vertical GHE were implemented in EnergyPlus. The GHE model also uses Eskilson's *f*-functions to model the response to time-varying heat fluxes and has been extended to include a computationally efficient variable time-step load aggregation scheme.

A vertical GHE model, based on the *f*-function algorithm, was implemented in the Building Creation Wizard of eQUEST. The eQUEST software is a very useful tool for GCHP system design and energy analysis. Both of the programs employ an effective steady-state borehole thermal resistance to calculate the actual heat transfer inside the borehole and have the flexibility to compare the energy consumption of the GCHP A/C system to a conventional heating, ventilating and air-conditioning system.

5.4.5.2 The GCHPCalc Program Based on the Cylindrical Source Model

The GCHPCalc is a program to help engineers with the design of vertical GCHP systems. The detailed fundamental concepts of this program can be found in Kavanaugh and Rafferty [34]. The method is based on a cylindrical source model and uses a simple steady-state heat transfer equation to solve the required borehole length (Eqn (5.5)), which considers three different 'pulses' of heat to account for long-term heat imbalances, average monthly heat rates during the design month and maximum heat rates for a short-term period during a design day. The thermal resistance of the ground responding to each pulse is calculated by means of modified solution of Carslaw and Jaeger [23].

5.4.5.3 Numerical Simulation Programs

Some numerical simulation codes, mainly based on the finite-difference method, have been developed in the GCHP field. Among them, the most representative numerical program is the Transient Systems Simulation (TRNSYS) [35] with a duct ground storage model (DST) module. The TRNSYS is a modular system simulation package where users can describe the components that compose the system and the manner in which these components are interconnected. Because the program is modular, the DST for vertical GHE, which does not take into account the grout thermal capacity is easily added to the existing component libraries.

5.4.6 Simulation of Ground Thermo-Physical Capacity

5.4.6.1 Ground Thermal Response Test

In the case of vertical closed-loop GCHP systems, the determination of the parameters to calculate the vapourisation thermal power that must be provided from the ground is laborious. Evaluating the thermal conductivity of the ground and the effective thermal resistance of the borehole are very important to know how many loops must be set, which is a function of the energy that must be given to the HP. In this respect, taking a thermal response test (TRT) of the ground is necessary, using a borehole in which a simple ground loop is placed.

Physical principles of the test. The thermal field surrounding the vertical GHE is determined with a line-source model, which represents the borehole as a heat line-source. Eqn (5.8) of the infinite line heat-source with constant intensity can be rewritten as [14]:

$$\Delta t(r_b, \tau) = t_b - t_g = \frac{q_E}{4\pi\lambda}\int_{r^2/4\cdot a\cdot\tau}^{\infty}\frac{e^{-u}}{u}du = \frac{q_E}{4\pi\lambda}\cdot E\left(\frac{r^2}{4a\tau}\right) \tag{5.22}$$

where $\Delta t(r_b, \tau)$ is the temperature difference dependent on the borehole radius r_b and the time τ; t_b is the average temperature of the borehole wall, in K; t_g is the undisturbed ground temperature, in K; q_E is the specific power of rejection/extraction, in W/m; λ is the thermal conductivity of the ground, in W/(m K); r is the effective radius, in m; $a = \lambda/\rho_c$ is the thermal diffusivity of the ground, in m^2/s; ρ is the ground density, in kg/m^3; c is the ground specific heat at constant pressure, in J/(kg K); and τ is the time, in s.

The exponential integral function E, for high values of the parameter $(a\tau/r^2)$, can be approximated with the following expression:

$$\mathrm{E}\left(\frac{r^2}{4a\tau}\right) = \ln\frac{4a\tau}{r^2} - \gamma \tag{5.23}$$

For $\tau > 5r_p^2/a$ Eqn (5.22) becomes:

$$\Delta t(r_p, \tau) = q_E R_g = \frac{q_E}{4\pi\lambda}\left(\ln\frac{4a\tau}{r_b^2} - \gamma\right) \tag{5.24}$$

where R_g is the thermal resistance of the ground, in K/(W/m) and γ is the Euler constant, approximately equal to 0.5772.

The temperature difference between the average temperature of the heat carrier fluid $t_f = (t_i + t_e)/2$ and the temperature on the borehole wall t_b (Figure 5.6), is given by:

$$t_f - t_b = R_b q_E \tag{5.25}$$

where R_b is the thermal resistance of the borehole, in K/(W/m).

By introducing the borehole thermal resistance R_b into Eqn (5.24), the temperature variation between the circulating fluid and the ground can be obtained:

$$\Delta t(r_b, \tau) = q_E(R_b + R_g) = q_E\left[R_b + \frac{1}{4\pi\lambda}\left(\ln\frac{4a\tau}{r_b^2} - \gamma\right)\right] \tag{5.26}$$

To obtain the smallest temperature differences in the borehole requires its thermal resistance to be as small as possible. This can be accomplished by increasing the ground thermal conductivity, using adequate filling materials and/or by increasing the distance between the tubes of the vertical loop.

Testing equipment. During an *in-situ* test, a ground electric heater usually provides heat to the circulating fluid (water or glycol) through the ground loop while the inlet (t_i) and outlet (t_e) fluid temperatures are measured (Figure 5.8). The average of these two instantaneous temperature readings is usually taken to represent the average temperature in the vertical ground loop at a given time. In an ideal test, the measured circulating flow rate and the heat input rate remain constant throughout the test. The first TRT in Romania was performed in 2009 by the GEOTHERM PDC company of Bucharest [36].

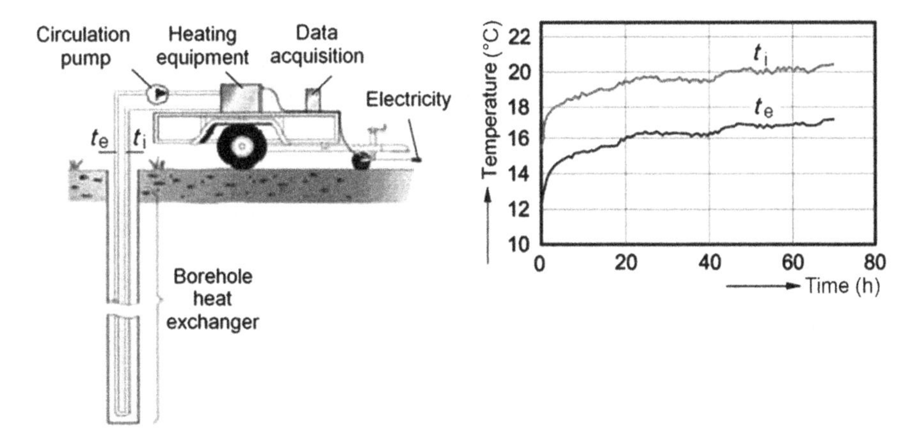

Figure 5.8 Schematic of an equipment for thermal response test.

Data analysis and final evaluation. For the amount of heat rejected/extracted from the ground, a transient state is set up, expressed by Eqn (5.24) in the form:

$$t_f = \frac{Q}{4\pi\lambda L}\ln(\tau) + \left[\frac{Q}{L}\left(\frac{1}{4\pi\tau}\left(\ln\frac{4a}{r_b^2} - \gamma\right) + R_b\right) + t_g\right] \tag{5.27}$$

where Q is the total power of rejection/extraction, in W and L is the borehole length, in m. Equation (5.27) can be simplified and written as line form:

$$t_f = \alpha \ln(\tau) + n \tag{5.28}$$

in which:

$$\alpha = \frac{Q}{4\pi\lambda L} \tag{5.29}$$

$$n = \frac{Q}{L}\left(\frac{1}{4\pi\lambda}\left(\ln\frac{4a}{r_b^2} - \gamma\right) + R_b\right) + t_g \tag{5.30}$$

where α is the late-time slope in a plot of the fluid temperature t_f versus the natural logarithm of time τ (Figure 5.9).

The ground thermal conductivity λ is obtained from Eqn (5.29):

$$\lambda = \frac{Q}{4\pi\alpha L} \tag{5.31}$$

The slope α of the interpolation straight-line of the measurements is independent of the borehole resistance R_b. Thus, an estimated thermal conductivity λ of the ground can be used to determine the real thermal resistance of the borehole. Figure 5.10 shows the variation of the average fluid temperature t_f depending on the time τ from the beginning of the test.

Replacing the ground thermal conductivity λ obtained from Eqn (5.31) with Eqn (5.24) results in the equivalent thermal resistance of the borehole:

$$R_b = \frac{1}{q_E}(t_f - t_b) = \frac{1}{q_E}(t_f - t_g) - \frac{1}{4\pi\lambda}\left(\ln(\tau) + \ln\frac{4a}{r_b^2} - \gamma\right) \tag{5.32}$$

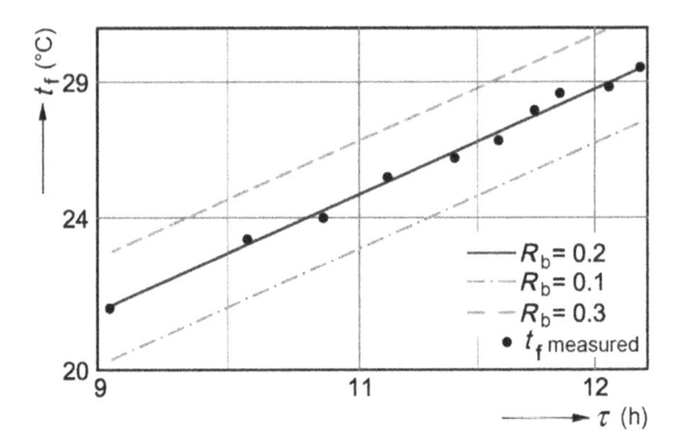

Figure 5.9 Determination of borehole thermal resistance.

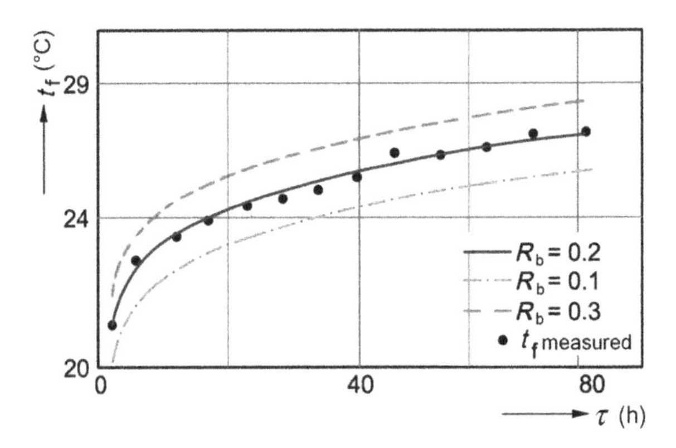

Figure 5.10 Variation of fluid temperature in time.

Equation (5.32) does not allow a proper estimation of the borehole equivalent thermal resistance, as it is influenced by the test duration through $\ln(\tau)$.

In addition, in Eqn (5.29), the thermal diffusivity comes in a, as a ratio between the thermal conductivity λ and the thermal capacity C. The thermal conductivity λ is determined by Eqn (5.31), but the thermal capacity C, in J/(m^3 K), can be only approximated using the equation:

$$C = \rho c = \exp\left[\ln\frac{4\tau\lambda}{r_b^2} - \gamma - \frac{4\pi\lambda}{q_E}(t_f - t_g - q_E R_b)\right] \tag{5.33}$$

where ρ is the ground density; c is the ground specific heat; and the borehole thermal resistance R_b is considered equal to 0.1 K/(W/m) for a standard borehole.

To estimate the minimum duration τ_{min} of the test, the following equation can be used [26]:

$$\tau_{min} = \frac{5r_b^2}{a} \tag{5.34}$$

where r_b is the borehole radius and a is the ground thermal diffusivity.

Austin et al. [37] recommend a minimum duration of 50 h based on their experience with field data sets. Gehlin [38] suggests a minimum duration of 60 h, but recommends using 72 h.

Bandyopadhyay et al. [39] obtained a semi-analytical solution for the short time transient response of a grouted borehole subjected to a constant internal heat generation rate.

For data analysis and final evaluation of ground thermal conductivity λ and borehole thermal resistance R_b, some methods were developed [13,37–41] that use one of the simulation models of GHEs previously presented. Through the ground TRT, the length of the borehole is properly determined, the operating performance of the system is provided and supplementary costs (e.g. extra loops, boreholes, glycol, etc.) are avoided. This operation is performed using specialised software.

5.4.6.2 Equivalent-Time for Interrupted Tests on BHE

Electricity outages, electric heater failures and other unexpected events sometimes interrupt borehole tests before the test duration is sufficient to estimate the ground thermal conductivity. Although it would be desirable to restart the test immediately after the equipment problems are fixed, the temperature distribution in the ground would have already changed. Most analysis methods assume a spatially uniform ground temperature at the start of the test, and this assumption is invalid if the test is restarted quickly.

To return to the initially undisturbed ground conditions, Martin and Kavanaugh [41] recommend a 10- to 12-day waiting period before retesting a borehole after a completed 48-h test. For an interrupted test, they suggest that the waiting period can be reduced in proportion to the reduced test time. Such time delays cost time and money when the equipment is on location and ready to restart the test.

If the duration of the interruption is no more than a few hours, the best course of action may be to restart the test immediately after the problem is repaired, even with standard analysis methods. As an example, Figure 5.11 illustrates the temperature rise curves from both uninterrupted and interrupted tests.

During the interrupted test, the electric energy was shut off for a 2-h period, starting approximately nine hours into the test.

The fluid temperature was greatly distorted immediately after start-up, but the rising temperature eventually overlays on the uninterrupted curve with nearly the same

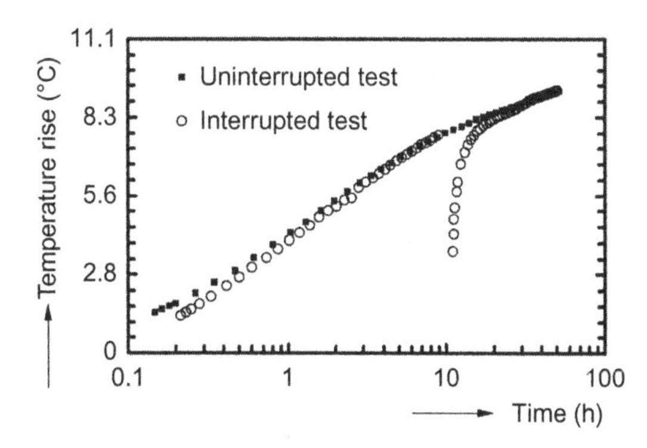

Figure 5.11 Fluid temperature curves.

late-time slope. The estimated thermal conductivity values from each test are within 2% of each other. In this case, the cumulative test time is 51 h, including the interruption. Thus, restarting the test immediately after the electric energy is restored is the best strategy in this case.

The equivalent-time method [42] for interrupted tests on BHEs rescales the time coordinate on the horizontal axis (Figure 5.11) and removes the effects of the interruption on the temperature rise curve. This method shortens the required test time for estimating the ground thermal conductivity.

Model for interrupted test. Considering the rate changes during an interrupted test, as shown in Figure 5.12, where the heat input rate is constant at the value of Q_1 between τ_1 and τ_2, but the electric energy supply is interrupted, the heat input rate suddenly goes to zero at time τ_1. Then, at time τ_2, when the electricity is restored, the heat input rate is restarted to Q_3, which may differ from the earlier value, Q_1. Superposition may be used to take into account these rate changes and to estimate the corresponding fluid temperature.

One applies the constant-rate solution for each step rate change ($Q_i - Q_{i-1}$), which occurs at time τ_{i-1}. If the number of rate changes is given as n, the fluid temperature is a sum of the constant-rate responses [40]:

$$t_f(\tau) = \sum_{i=1}^{n} \frac{Q_i - Q_{i-1}}{Q_{\text{ref}}} t_u(\tau - \tau_{i-1}) \tag{5.35}$$

where $\tau_{n-1} \leq \tau \leq \tau_n$ and $Q_0 = 0$ at τ_0. For the line-source model, t_u is set equal to t_f in Eqn (5.27), where Q is set equal to $(Q_i - Q_{i-1})$ for each step change. The reference heat input rate is set to the last input rate change, $Q_n - Q_{n-1}$. For the rate schedule in Figure 5.12, the reference heat input rate is $Q_3 - Q_2$.

Equivalent-time method. Because the interruption in the heat input rate greatly distorts the temperature rise in Figure 5.11, an alternative to the conventional line-source analysis is needed. This is the motivation behind the equivalent time method. For a single interruption, the equivalent time is given by [40]:

$$\Delta\tau_e = \left(\frac{\tau_2 + \Delta\tau}{\tau_2}\right)^{\frac{Q_1}{Q_3 - Q_2}} \left(\frac{\tau_2 - \tau_1 + \Delta\tau}{\tau_2 - \tau_1}\right)^{\frac{Q_2 - Q_1}{Q_3 - Q_2}} \Delta\tau \tag{5.36}$$

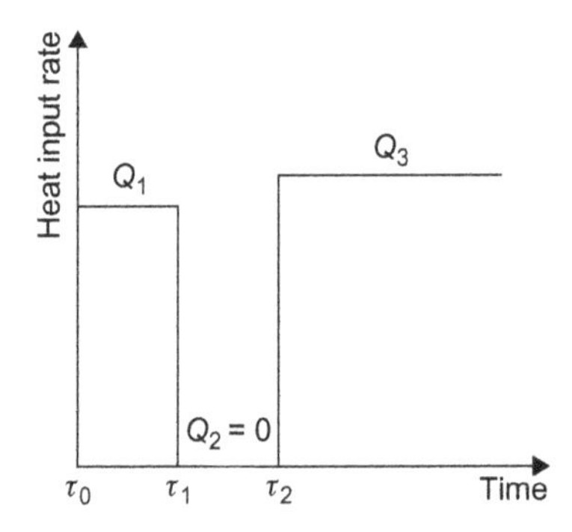

Figure 5.12 Variations in heat input rate represented as discrete step changes during an interrupted test.

where $\Delta\tau = \tau - \tau_2$ and $\tau_2 < \tau$. The temperature rise after the interruption may be expressed in terms of equivalent time as [40]:

$$t_f - t_2 = \frac{Q_3 - Q_2}{4\pi\lambda L}\left[\ln\left(\frac{4a\Delta\tau_e}{\gamma r_b^2}\right) + 4\pi\lambda R_b\right] \tag{5.37}$$

Thus, equivalent time $\Delta\tau_e$ transforms the temperature rise in Eqn (5.37) into the same mathematical formulation as the constant heat input rate case in Eqn (5.27). Equivalent time takes the place of the elapsed test time τ and t_2 takes the place of the undisturbed ground temperature t_g.

A comparison of Eqns (5.27) and (5.37) shows that the expression for ground thermal conductivity for the equivalent time method is a simple modification of Eqn (5.31):

$$\lambda = \frac{Q_3 - Q_2}{4\pi\alpha L} \tag{5.38}$$

The slope α in Eqn (5.38) comes from a semilog graph with the natural logarithm of equivalent time on the horizontal axis.

Similarly, the expression for the borehole thermal resistance based on the equivalent time, after some algebraic rearrangement of Eqn (5.37), gives:

$$R_b = \frac{1}{4\pi\lambda}\left[\frac{t_{f,1\,h} - t_2}{\alpha}\right] - \ln\left(\frac{4a\Delta\tau_{e,1\,h}}{\gamma r_b^2}\right) \tag{5.39}$$

where the slope α is from a graph with the natural logarithm of equivalent time on the horizontal axis.

If the number of heat input rates in a field test is more than three, the same techniques would apply. For the application with n heat input rates, the term $(Q_3 - Q_2)$ in Eqn (5.38) is replaced by $(Q_n - Q_{n-1})$ to estimate the ground thermal conductivity. Likewise, the term t_2 is replaced by t_{n-1} in Eqn (5.39) to estimate the borehole resistance.

The equivalent time method differs from the practice of waiting for the temperature distribution in the ground to approach its undisturbed uniform temperature, which may require a long delay before restarting the test [41]. Such delays cost money if the equipment is repaired and waiting for resumption of the test. The equivalent time method allows an analysis of test data if the test is restarted as soon as possible. The minimum additional time after restart is less than or equal to the minimum test duration for an uninterrupted test. Therefore, if one takes into account the interrupted heat input rates with equivalent time and restarts the test as soon possible, the additional test time is no more than for the case of starting a test with undisturbed ground temperature.

5.4.6.3 Use of EED Simulation Program

- *The EED program input data* are as follows:
  - **Ground**:
    - Thermal conductivity: 1.90 W/(m K)
    - Thermal capacity: 2.40 MJ/(m^3 K)
    - Annual average temperature of the ground surface: 10.6 °C
    - Geothermal heat flux: 0.07 W/m^2

- **Borehole**:
  - Number of boreholes: 1
  - Configuration: 0 ('1:single')
  - Borehole length: 80.00 m
  - Collector type: Single-U
  - Borehole diameter: 110 mm
  - U-tube diameter: 32 mm
  - U-tube wall thickness: 3 mm
  - U-tube thermal conductivity: 0.42 W/(m K)
  - Distance between axes of the tubes: 60 mm
  - Thermal conductivity of the filler: 0.60 W/(m K)
  - Contact resistance filler-tube: 0.00 (m K)/W
- **Circulating fluid**:
  - Thermal conductivity: 0.480 W/(m K)
  - Specific heat: 3795 J/(kg K)
  - Density: 1052 kg/m^3
  - Viscosity: 0.0052 kg/(m s)
  - Freezing temperature: − 14 °C
  - Flow rate through the probe: 0.300 l/s
- **The basic load**:

  Seasonal performance factor for:

  heating: 4.00; cooling: 99999.0; and DHW: 3.70.

  The values of the monthly thermal energy demand are given in Table 5.9.
- **The peak load:**
  - The peak monthly thermal loads are given in Table 5.10.
  - Simulation years: 25
  - First month of operation: April

Table 5.9 Monthly Thermal Energy Demands, in MWh

Month	Heating Load	Cooling Load	Ground Thermal Load
January	0.89	0.00	0.657
February	0.75	0.00	0.556
March	0.67	0.00	0.492
April	0.45	0.00	0.331
May	0.35	0.25	0.001
June	0.35	0.35	− 0.099
July	0.35	0.57	− 0.312
August	0.35	0.51	− 0.254
September	0.35	0.25	0.007
October	0.54	0.00	0.399
November	0.70	0.00	0.514
December	0.83	0.00	0.616
Total	6.56	1.93	2.908

Table 5.10 Peak Monthly Thermal Loads, in kW

Month	Peak Heating Load	Duration (h)	Peak Cooling Load	Duration (h)
January	3.11	24.0	0.00	0.0
February	3.11	24.0	0.00	0.0
March	0.00	0.0	0.00	0.0
April	0.00	0.0	0.00	0.0
May	0.00	0.0	0.00	0.0
June	0.00	0.0	2.15	10.0
July	0.00	0.0	2.15	10.0
August	0.00	0.0	2.15	10.0
September	0.00	0.0	0.00	0.0
October	0.00	0.0	0.00	0.0
November	0.00	0.0	0.00	0.0
December	3.11	24.0	0.00	0.0

- *Simulation results.* The results obtained using the EED program are as follows:
  - **Thermal resistances**:
    - Borehole thermal resistance: 0.7141 m K/W
    - Reynolds number: 2972
    - Fluid-tube thermal resistance: 0.0139 m K/W
    - Tube thermal resistance: 0.0787 m K/W
    - Grout-tube thermal resistance: 0.0000 m K/W
    - Fluid-ground thermal resistance: 0.1856 m K/W
    - Effective borehole thermal resistance: 0.1877 m K/W
  - **Specific extraction power**:

    Table 5.11 presents the monthly values of the specific extraction power for basic and peak heating and cooling.
  - **The average fluid temperatures**:

Table 5.12 summarises the monthly average temperatures for the basic thermal load, depending on the year of the simulation. In the 25th simulation year, the minimum average fluid temperature was 5.52 °C in late January, and the maximum average fluid temperature was 15.82 °C at the end of July.

Table 5.13 gives the monthly average fluid temperatures for the peak heating load, depending on the year of the simulation. In the 25th simulation year, the minimum average fluid temperature was −0.79 °C in late January, and the maximum average fluid temperature was 13.99 °C is at the end of July.

Table 5.14 gives the monthly average fluid temperatures for the peak cooling load, depending on the year of the simulation. In the 25th simulation year, the minimum average fluid temperature was 5.52 °C in late January, and the maximum average fluid temperature was 20.80 °C at the end of August.

Table 5.11 Specific Extraction Power, in W/m

Month	Basic Load	Peak Heating Load	Peak Cooling Load
January	11.24	29.16	0.00
February	9.52	29.16	0.00
March	8.42	0.00	0.00
April	5.67	0.00	0.00
May	0.02	0.00	0.00
June	− 1.70	0.00	− 26.88
July	− 5.34	0.00	− 26.88
August	− 4.35	0.00	− 26.88
September	0.12	0.00	0.00
October	6.83	0.00	0.00
November	8.80	0.00	0.00
December	10.55	29.16	0.00

Table 5.12 The Average Fluid Temperatures for the Basic Load, in °C

Simulation Year	1	2	5	10	25
January	12.07	6.02	5.75	5.63	5.52
February	12.07	6.74	6.48	6.36	6.26
March	12.07	7.23	6.99	6.87	6.77
April	9.26	8.57	8.34	8.23	8.12
May	11.90	11.42	11.20	11.09	10.99
June	12.82	12.44	12.23	12.13	12.02
July	14.70	14.39	14.19	14.09	13.99
August	14.36	14.09	13.90	13.80	13.70
September	12.20	11.97	11.79	11.69	11.59
October	8.80	8.59	8.42	8.32	8.22
November	7.59	7.41	7.24	7.14	7.04
December	6.54	6.37	6.21	6.11	6.01

Figure 5.13 plots the evolution of the average fluid temperature in the ground for cooling, heating and DHW production thermal power over a period of 25 years. Four different scenarios were simulated, two for the winter season (base load and peak load for heating) and two for the summer season (base load and peak load for cooling).

Analysing the time evolution of the fluid temperatures in the ground for the peak loads reveals that these values are approximately constant, meaning that the heat source (ground) is fully regenerated and thus the GCHP will maintain high performance in operation.

Table 5.13 The Average Fluid Temperatures for the Peak Heating Load, in °C

Simulation Year	1	2	5	10	25
January	12.07	− 0.28	− 0.56	− 0.68	− 0.79
February	12.07	− 0.18	− 0.44	− 0.55	− 0.66
March	12.07	7.23	6.99	6.87	6.77
April	9.26	8.57	8.34	8.23	8.12
May	11.90	11.42	11.20	11.09	10.99
June	12.82	12.44	12.23	12.13	12.02
July	14.70	14.39	14.19	14.09	13.99
August	14.36	14.09	13.90	13.80	13.70
September	12.20	11.97	11.79	11.69	11.59
October	8.80	8.59	8.42	8.32	8.22
November	7.59	7.41	7.24	7.14	7.04
December	− 0.01	− 0.18	− 0.34	− 0.44	− 0.54

Table 5.14 The Average Fluid Temperatures for the Peak Cooling Load, in °C

Simulation Year	1	2	5	10	25
January	12.07	6.02	5.75	5.63	5.52
February	12.07	6.74	6.48	6.36	6.26
March	12.07	7.23	6.99	6.87	6.77
April	9.26	8.57	8.34	8.23	8.12
May	11.90	11.42	11.20	11.09	10.99
June	20.76	20.38	20.18	20.07	19.97
July	21.50	21.18	20.99	20.88	20.78
August	21.47	21.20	21.01	20.91	20.80
September	12.20	11.97	11.79	11.69	11.59
October	8.80	8.59	8.42	8.32	8.22
November	7.59	7.41	7.24	7.14	7.04
December	6.54	6.37	6.21	6.11	6.01

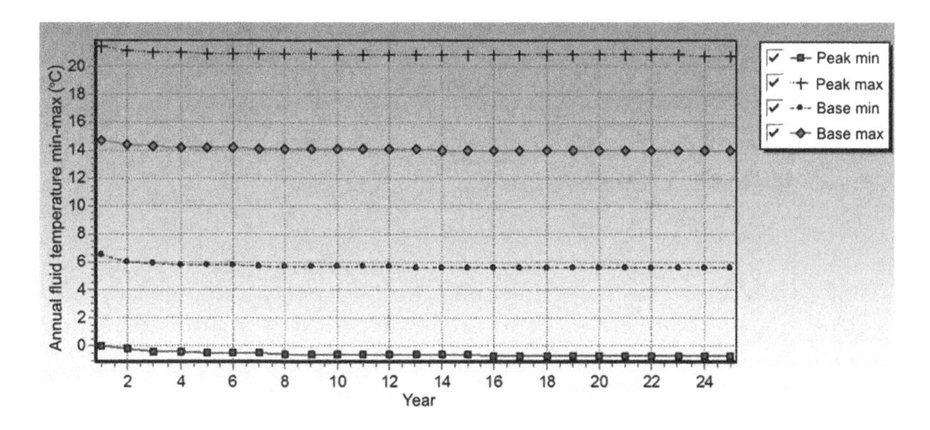

Figure 5.13 Average fluid temperature in the ground over a period of 25 years.

5.4.7 One-Dimensional Transient BHE Model

Shirazi and Bernier [43] developed a one-dimensional transient BHE model to account for fluid and grout thermal capacities in BHEs to predict the outlet fluid temperature for varying inlet temperature and flow rate. The borehole model is coupled with a ground model which is based on the cylindrical heat source method. Using this model in energy simulation programs can quantify the effect of borehole thermal capacity on HP annual energy consumption.

5.4.7.1 Model Description

The following analysis is based on the assumption that the essence of transient heat transfer in boreholes can be captured by replacing the U-tube, two-pipe geometry, by a grout-filled cylinder delimited by an inside equivalent diameter and the real borehole diameter. The transformation of the geometry is illustrated in Figures 5.14 and 5.15.

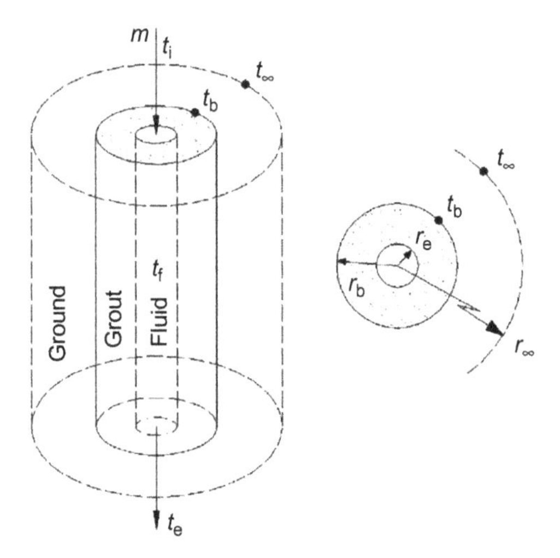

Figure 5.14 Transformation from two-pipe geometry to an equivalent single pipe.

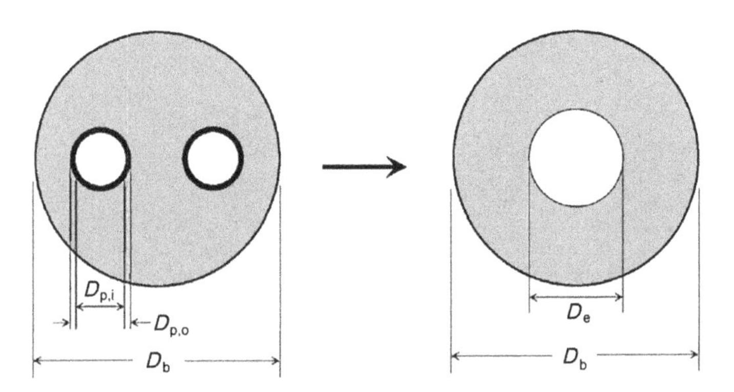

Figure 5.15 Approximation of real geometry by a cylinder core with an equivalent inside diameter.

The resulting one-dimensional approximation neglects the axial (along the length of the borehole) and azimuthally (along the circumference) variations and only considers radial variations. Even though this approach does not provide the fine details that a 3D transient model would give, it has the advantage of being computationally less intensive which enables inclusion of borehole (fluid and grout) thermal capacities in annual energy simulations.

The objective is to predict the outlet fluid temperature for a given set of transient fluid temperature and flow rate at the borehole inlet. Transient radial heat transfer in the cylinder core (comprised of the grout and the fluid) is solved numerically, while the ground outside the borehole is treated analytically using the cylindrical heat source method. It would have been possible to extend the numerical calculation domain to include the ground and this approach would have worked adequately for single boreholes. It was decided to separate the two domains so that the borehole portion of the computation model could eventually be coupled to *f*-function based ground model which could handle multiple boreholes. Furthermore, this approach reduces the required computational time. These two models are coupled through the heat flux and temperature at the borehole diameter. The temperature variation in the fluid is obtained from an energy balance. Since the fluid temperature is unknown a priori, an iterative procedure is used to obtain the outlet fluid temperature.

Equivalent pipe diameter approximation. As shown in Figure 5.15, the standard U-tube borehole is approximated by a core cylinder with an equivalent inside diameter and outside diameter corresponding to the real borehole diameter. The thermal capacity of the HDPE pipes is neglected. The comparison presented in Table 5.15 (based on: two SDR-11 pipes, borehole diameter: 0.15 m, volumetric heat capacities for pipe and grout: 1.77 and 3.9 MJ/(m^3 K), respectively) shows that this assumption is justified because the grout thermal capacity is usually much higher than the pipe thermal capacity.

The equivalent pipe diameter D_e is based on the steady-state borehole thermal resistance $R_{b,ss}$ of Hellström [32]:

$$R_{b,ss} = \frac{1}{4\pi\lambda_{gt}}\left[\ln\left(\frac{r_b}{r_p}\right) + \ln\left(\frac{r_b}{2D}\right) + \sigma \ln\left(\frac{r_b^4}{r_b^4 - D^4}\right)\right] + \frac{1}{2}R_p \tag{5.40}$$

in which:

$$\sigma = \frac{\lambda_{gt} - \lambda}{\lambda_{gt} + \lambda'} \quad R_p = \frac{\ln(D_{p,o}/D_{p,i})}{2\pi\lambda_p} \tag{5.41}$$

Table 5.15 Comparison Between Pipe and Grout Thermal Capacities in Typical Boreholes

HIDPE Pipes (mm)	Thermal Capacity per Unit Length of Borehole (J/(m K))		Thermal Capacity Ratio
	Pipe	Grout	Grout/pipe
19	325	64,600	199
25	505	62,100	123
30	815	58,000	71

where: D is half the centre-to-centre distance between the two legs of the U-tube; λ and λ_{gt} are the ground and grout thermal conductivities; r_b is the borehole radius; $r_p = D_{p,o}/2$ is the pipe radius; and R_p is the pipe thermal resistance. The equivalent diameter D_e is obtained by equating the steady-state borehole thermal resistance for the real geometry $R_{b,ss}$ to the thermal resistance of the equivalent cylinder:

$$R_{b,ss} = \frac{\ln(D_b/D_e)}{2\pi\lambda_{gt}} \tag{5.42}$$

where D_b is the borehole diameter. When solving for the equivalent pipe diameter D_e, one obtains the formula:

$$D_e = D_b \exp(-2\pi\lambda_{gt} R_{b,ss}) \tag{5.43}$$

To account for possible flow rate variations during simulations, the thermal resistance associated with the internal film coefficient, which is not included in $R_{b,ss}$, is calculated separately and added to $R_{b,ss}$.

The transformation from U-tube geometry to an equivalent cylinder has repercussions on the fluid velocity in the borehole and the internal film coefficient. Modifications are therefore necessary in the equivalent geometry to maintain identical heat transfer in both geometries. The first of these modifications concerns the equivalent internal film coefficient h_e, which is given by:

$$h_e = 2\frac{D_{p,i}}{D_e}h \tag{5.44}$$

where h is the internal film coefficient of the U-tube configuration. The value can be obtained from standard correlations. The residence time of the fluid in the BHE, t_{res} corresponds to time required for the fluid to travel from the inlet to the outlet. For the U-tube configuration t_{res} is given by:

$$t_{res} = \rho_f A \frac{2L}{m} \tag{5.45}$$

where ρ_f is the fluid density; $A = \pi D_{p,i}^2/4$ is the cross-sectional area of the U-tube pipe, and m is the mass flow rate. For the equivalent cylinder, the residence time is:

$$t_{res,e} = \rho_{f,e} A_e \frac{L}{m} \tag{5.46}$$

where 'e' refers to the equivalent cylinder geometry and $A_e = \pi D_e^2/4$ is the cross-sectional area of the equivalent cylinder. The mass flow rates and residence times must be identical in both the real and equivalent geometries, which imply that the fluid density ($\rho_{f,e}$) and velocity (u_e) in the equivalent cylinder are equal to:

$$\rho_{f,e} = 2\rho_f \left(\frac{D_{p,i}}{D_e}\right)^2 \tag{5.47}$$

$$u_e = \frac{\rho_f}{\rho_{f,e}} \left(\frac{D_{p,i}}{D_e}\right)^2 u \tag{5.48}$$

where u is the fluid velocity in the U-tube configuration.

From Eqns (5.47) and (5.48):

$$u_e = 0.5u \tag{5.49}$$

Finally, the total fluid thermal capacity inside the equivalent pipe has to be identical to that of the real geometry. This leads to:

$$c_{f,e} = 2\left(\frac{D_{p,i}}{D_e}\right)^2 \frac{\rho_f}{\rho_{f,e}} c_f \tag{5.50}$$

where c_f and $c_{f,e}$ are the specific heats of the fluid in the U-tube geometry and in the equivalent pipe, respectively. From Eqns (5.47) and (5.50):

$$c_{f,e} = c_f \tag{5.51}$$

To summarise, based on the known characteristics of the U-tube geometry (i.e. D_b, λ_{gt}, $D_{p,i}$, L) and the operating conditions (i.e. h, m, ρ_f, c_f) Eqns (5.43), (5.44), (5.47), (5.49) and (5.51) are used to obtain the corresponding values for the equivalent geometry.

Transient heat transfer in the borehole. Transient heat transfer in the equivalent geometry is solved numerically. The problem is governed by the unsteady one-dimensional energy equation in cylindrical coordinates:

$$\rho_{gt} c_{p_{gt}} \frac{\partial t}{\partial \tau} = \lambda_{gt} \frac{1}{r} \frac{\partial}{\partial r}\left(r \frac{\partial t}{\partial r}\right) \tag{5.52}$$

where the subscript 'gt' refers to grout properties which are assumed to be constant. Figure 5.16 presents the extent of the calculation domain. Equation (5.52) is subject to the following boundary conditions:

$$\begin{aligned} t &= t_b(\tau) \quad \text{at } r = r_b \\ -2\pi r_e \lambda_{gt} \frac{\partial t}{\partial r}\Big|_{r=r_e} &= q'(\tau) \\ q'(\tau) &= 2\pi r_e h_e (t_{r_e} - t_f) \end{aligned} \tag{5.53}$$

where $r_e = D_e/2$ is the equivalent radius; t_{r_e} is the temperature at the equivalent radius; and $t_f = (t_i + t_e)/2$ is the mean fluid temperature in the borehole. The whole domain is assumed to be at the undisturbed ground temperature, t_∞ at $\tau = 0$.

The problem is solved using the control volume-based finite difference method with the fully implicit approach. Using the nomenclature presented in Figure 5.16, the discretised equation for an internal node P is given by:

$$a_P T_P = a_N T_N + a_s T_s + b \tag{5.54}$$

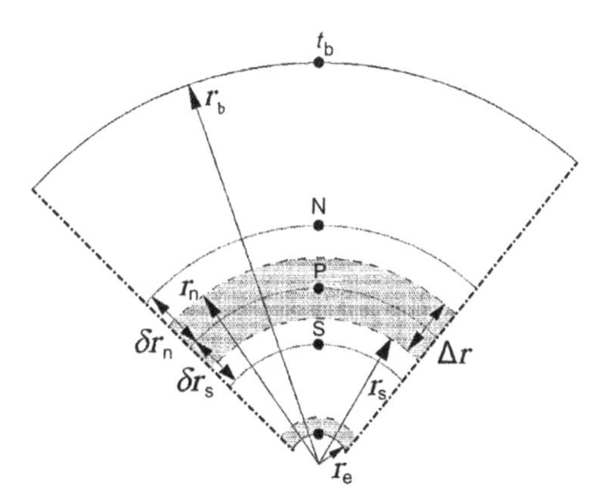

Figure 5.16 Grids in the calculation domain of the equivalent geometry.

where the coefficients are:

$$a_N = \frac{r_n \lambda_{gt,n}}{\delta r_n}, \quad a_S = \frac{r_s \lambda_{gt,s}}{\delta r_s}, \quad a_P^0 = \frac{\rho_{gt} c_{gt}}{2\Delta\tau}(r_n^2 - r_s^2),$$
$$b = a_P^0 t_P^0, \quad a_P = a_P^0 + a_N + a_S \tag{5.55}$$

The superscript '0' refers to the conditions at the previous time step, and $\Delta\tau$ is the time step. The first node is located on the interior wall in a half control volume. For this particular node, the discretised equation for the temperature t_{r_e} at r_e is given by:

$$a_P t_{r_e} = a_N t_N + b \tag{5.56}$$

where:

$$a_N = \frac{r_n \lambda_{gt,n}}{\delta r_n}, \quad a_P^0 = \frac{\rho_{gt} c_{gt}}{2\Delta\tau}(r_n^2 - r_e^2) \tag{5.57}$$

In Eqn (5.56) when heat is transferred from the fluid to the pipe, then $q''_{r_e} = h_e(t_f - t_{r_e})$.

A non-uniform grid structure was used with a resulting concentration of grid points near step temperature gradients. The results of the time and grid independence checks indicated that 10 grid points are needed in the core cylinder and that the time step should be approximately equal to RT 20 (where RT is the residence time) in order to obtain a solution that is independent of the time step and grid spacing.

Treatment of the fluid thermal capacity. The one-dimensional approach assumed in the developed model implies that fluid axial conduction is neglected and that the mean fluid temperature t_f is uniform over the full length of the borehole during a given time step. Based on these assumptions, and assuming constant fluid thermal properties, a transient energy balance on the fluid volume in the borehole yields:

$$mc_{f,e}(t_i - t_e) - h_e A_e(t_f - t_{r_e}) = \rho_{f,e} c_{f,e} V_e \frac{dt_f}{d\tau} \tag{5.58}$$

where V_e is the fluid volume in the borehole. Discretising Eqn (5.58) using the implicit approach gives:

$$\dot{m}c_{f,e}(t_i - t_e) - 2\pi r_e L h_e(t_f - t_{r_e}) = \frac{\rho_{f,e} c_{f,e} \pi r_e^2 L(t_f - t_f^0)}{\text{RT}} \tag{5.59}$$

Equation (5.59) states that the net energy entering the control volume by advection minus the energy leaving it by convection is equal to the fluid internal energy variation over a certain period of time, which in this case is the fluid residence time, RT. On the left hand side of Eqn (5.59) all temperatures are evaluated at the current time step while on the right hand side, t_f^0 is the mean fluid temperature prevailing during the previous time step.

Substituting for t_f^0 and rearranging Eqn (5.59) to solve for t_e at the current time step, yields:

$$t_e = \frac{1}{B}\left[Ct_i + Dt_{r_e} + A(t_e^0 + t_i^0)\right] \tag{5.60}$$

where:

$$A = \frac{\rho_{f,e} c_{f,e} \pi r_e^2 L}{2\text{RT}}, \quad B = mc_{f,e} + \pi r_e L h_e + A_e,$$
$$C = mc_{f,e} - \pi r_e L h_e - A_e, \quad D = 2\pi r_e L h_e \tag{5.61}$$

When there is no flow in the borehole, Eqn (5.59) reduces to:

$$t_f = \frac{Et_f^0 + t_{r_e}}{1+E}, \quad \text{where } E = \frac{\rho_{f,e} c_{f,e} r_e}{2h_e \text{RT}} \tag{5.62}$$

When fluid thermal capacity is neglected, Eqn (5.62) reduces to $t_f = t_{r_e}$. In addition, if the grout thermal capacity is neglected, then t_f and t_{r_e} are equal to the borehole wall temperature t_b. Thus, when there is no flow and the grout and fluid thermal capacities are neglected, the fluid temperature is equal to the borehole wall temperature.

Heat transfer in the ground. Heat transfer from the borehole wall to the far-field is similar to the heat transfer from a cylinder subjected to a heat flux boundary condition on its inner wall and embedded in an infinite homogeneous medium with a constant far-field temperature, t_∞. For a constant heat flux boundary condition, the time variation of the borehole wall temperature $t_{b,\tau}$ given by the cylindrical heat source is:

$$t_{b,\tau} = t_\infty - \frac{q}{L}\frac{G(\text{Fo})}{\lambda} \tag{5.63}$$

where q is the heat transfer rate (a positive 'q' value implies heating, i.e. heat transfer from the ground to the fluid); $G(\text{Fo})$ is the solution indicated by Cooper [44] for a given Fourier number (Fo) defined as $\text{Fo} = a\tau/r_b^2$, in which a is the ground thermal diffusivity and τ is the time.

When performing energy simulations the heat flux will vary. In these cases, the temporal superposition and load aggregation algorithm is used. The technique, referred to as the Multiple Load Aggregation Algorithm (MLAA), uses two major thermal history periods, referred to as 'past' and 'immediate'. This technique leads to:

$$t_{b,\tau} = t_\infty - \frac{1}{\lambda L}(\text{MLAA}) \tag{5.64}$$

where MLAA represents the terms of the multiple load aggregation algorithm [45].

5.4.7.2 Solution Methodology

The solution flow chart for the described model is given in Figure 5.17. As shown in this figure, for each time increment TI, there are two inner iterative loops identified as 'numerical borehole model' and 'intermediate calculations'. Several temporal variables are used. The time increment (TI) is the time difference between two step changes in the inlet conditions. The value of TI can vary from a few minutes to 1 h in energy simulation programs. The residence time RT given by Eqn (5.45) is the time required for the fluid to travel through the GHE and ranges from 2 to 5 min in full flow. A number of intermediate calculations (IT) are performed during each TI. The value of IT is equal to the truncated value of the ratio TI/RT. During each RT, a number of calculations (NI) are performed in the numerical borehole model. In other words, the time step in the numerical simulations, $d\tau$, is given by the ratio RT/NI.

The main objective of the calculations is the determination of the outlet fluid temperature from the borehole t_e, for values of t_i and m prevailing during a given TI. The procedure starts by calculating the equivalent parameters given by Eqns (5.43), (5.44), (5.47), (5.49) and (5.51). Then, for the current time increment TI, t_f^0 and t_b are set as equal to their corresponding values at the previous TI. The intermediate calculations loop is initiated with a guess for t_f for the numerical borehole model.

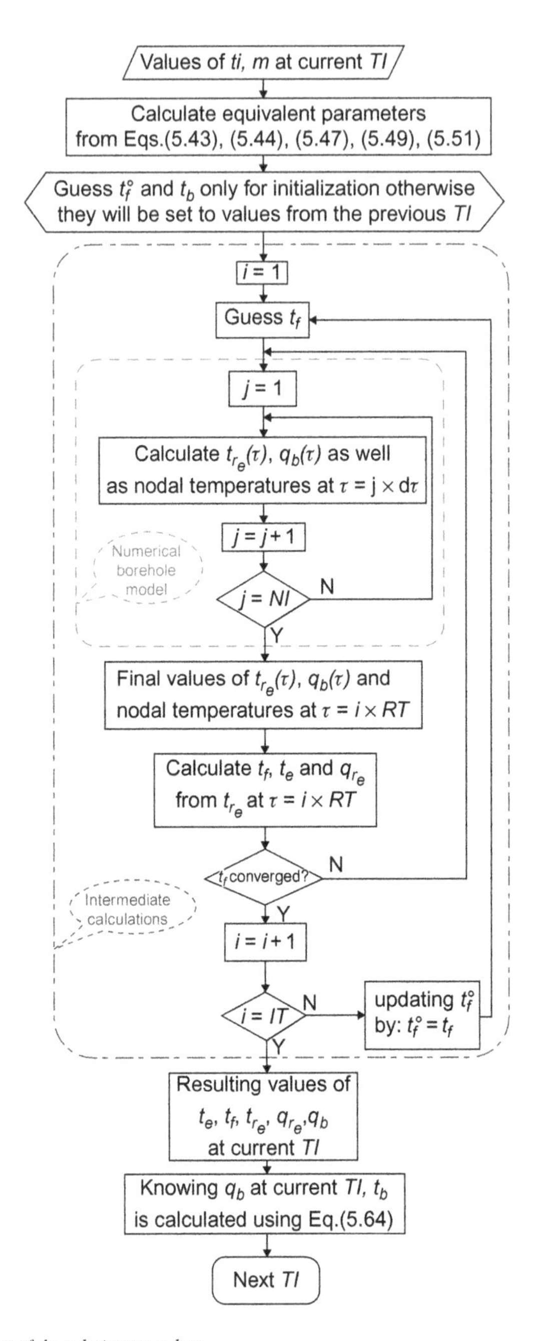

Figure 5.17 Flow chart of the solution procedure.

Then, t_{r_e} and q_b as well as the nodal temperatures in the equivalent cylinder are determined for each $d\tau$. At the last NI (an amount of time equivalent to RT has passed) calculations are performed in the numerical borehole model before passing the values of t_{r_e} and q_b to the intermediate calculations loop.

The values of t_{r_e} and q_b are then used calculate t_f, t_e and q_{r_e} at a time corresponding to the product $i \times$ RT. Then, the recently calculated value of t_f is compared to the guessed value of t_f. If both values do not agree within a certain tolerance, calculations are repeated with a new value of t_f. At convergence, t_f^0 is set to the recent value of t_f and the next intermediate calculation starts, intermediate calculations proceed until $i =$ IT. Then, values of t_e, t_f, t_{r_e}, q_{r_e} and q_b at the current TI are calculated. Finally, before returning to the next TI, t_f is evaluated using Eqn (5.64) from the calculated value of q_b at the current TI.

The described model was validated successfully against analytical solutions, experimental data, a three-dimensional transient numerical model and TRNSYS's Type 451. This model can be used to study the impact of borehole thermal capacity on the prediction of the outlet fluid temperature. Results show that the outlet fluid temperature is always higher when borehole thermal capacity is included [43].

5.4.8 Modelling the Interactions Between Ground Temperature Variations and Performance of GCHPs

A GSHP typically uses the ground as a heat source during the heating season and a heat sink during the cooling season. Because of the relatively constant underground temperature (higher than the air temperature during winter and lower than the air temperature during summer), the GSHP theoretically has higher energy efficiency compared with the conventional air-to-air or air-to-water HPs. The GHE is an important part of the GSHP system.

The accurate prediction of long term ground temperature distribution when the GSHP is running can improve the design of the GHE. Previously, some models were developed to predict the ground temperature distribution, such as the analytical solution of line source theory, the analytical solution of cylindrical source theory and numerical solutions. Analytical solutions can deliver accurate results in a timely manner. However, they only suit single boreholes. Numerical simulation methods are widely used in engineering projects to investigate long-term temperature distribution of ground for multiple boreholes and predict the HP performance. Qian and Wang [46] developed a model to integrate both the GCHP system performance and the ground temperature distribution.

5.4.8.1 Model Description

Multiple boreholes were placed as array as shown in Figure 5.18. Boreholes were placed at the nodes of the rectangle mesh. The distance between boreholes was between 3 and 6 m. There are a number of tube configurations (e.g. U-tube and double U-tube). Each borehole is simplified as one single point as the heat source. The difference between different configurations is expressed by the different heat resistance shown as R_g in Eqn (5.69). Figure 5.18 shows the layout of boreholes in two dimensions of 5×5 multiple pipes.

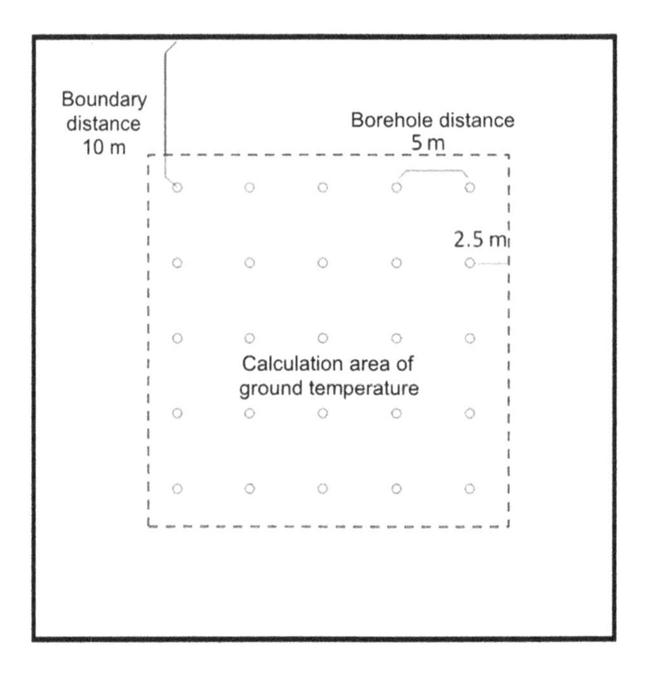

Figure 5.18 Computation area and array of boreholes.

The heat transfer around a vertical multi-pipe in ground is governed by soil thermal properties, distance between boreholes, cooling and heating loads and outdoor air temperature. The following assumptions are considered:

1. The initial ground temperature is uniform.
2. The ground properties are constant.
3. The influence of outdoor air temperature and sub-surface seepage flow is negligible because the impact of outdoor air temperature is insignificant for vertical pipes [19].
4. The vertical heat transfer is excluded. The problem can be simplified as two dimensions.
5. The temperature of fluid for a vertical pipe is constant throughout the entire length of pipe.

Based on the above assumptions, the governing equation is described as:

$$\frac{\partial t_g}{\partial \tau} = a\left(\frac{\partial^2 t_g}{\partial x^2} + \frac{\partial^2 t_g}{\partial y^2}\right) + \frac{q}{\rho c} \tag{5.65}$$

where: t_g is the ground temperature, in °C; τ is time, in s; a is the ground thermal diffusivity, in m^2/s; ρ is the ground density, in kg/m^3; c is the specific heat capacity, in J/(kg °C); q is the heat source, in W/m^3.

Initial condition: when $\tau = 0$, $t(x,y,\tau) = t_0$, where t_0 is the initial ground temperature, in °C, and τ is time, in s.

Equation (5.65) can be discretised in space and time. The implicit algorithm was employed to discretise time to make results insensitive to time step.

Boundary conditions:

1. The constant temperature was set as the measured ground temperature at the boundary of computing domain (Figure 5.18). The distance between the outside borehole and the computing boundary is 10 m, which is twice the distance between boreholes.
2. At the borehole node, the heat source q (W/m^3) was set to $\varphi/(\Delta x \Delta y)$, where $\Delta x \Delta y$ is the grid distance of the node in which the borehole is located (both Δx and $\Delta y = 0.1$ m), and φ is set as the heat exchange per metre between circulating fluid and ground, in W/m. At other locations, q is set to 0. It can be obtained based on the cooling or heating load and the performance of GSHP shown in Eqns (5.66) and (5.67).

For cooling load:

$$\varphi = \left[Q_{\mathrm{rac}}\left(1 + \frac{1}{\mathrm{COP}}\right)\right]/L \tag{5.66}$$

For heating load:

$$\varphi = \left[Q_{\mathrm{inc}}\left(1 - \frac{1}{\mathrm{COP}}\right)\right]/L \tag{5.67}$$

where: Q_{rac} is the cooling load, in W, in the summer; Q_{inc} is the heating load, in W, in the winter; L is total length of boreholes, in m; COP is the coefficient of performance.

Different HPs, different condensation temperatures and evaporation conditions lead to different COP values. HP manufactures normally publish based on a single COP value for one model under standard conditions. Some manufactures may publish several COP values for one model under various conditions of supply/return chilled and cooling water temperature. The industry-average data can be calculated using the following equation [47]:

$$\mathrm{COP_{GSHP}} = 0.000734\Delta t^2 - 0.150\Delta t + 8.77 \tag{5.68}$$

where Δt is the difference between the average water temperature in the HPs condenser and the evaporator. The average water temperature was simplified as the average water temperature of entering and exiting. The difference Δt should be between 20 and 60 °C.

The default supplied/returned chilled water to/in the building temperature is 7/12 °C in summer, which means the average source temperature in summer is 9.5 °C. The supplied/returned heated water temperature distributed to the building is 45/40 °C in winter, which means average sink water temperature in winter is 42.5 °C. The COP value of GCHP depends on the water (circulating fluid) from the GHE. The average sink water temperature in the summer and the source water temperature in the winter can be calculated based on the average temperature t_f of water going into and going out of the GHE ($t_f = (t_i + t_e)/2$), which can be obtained by conservation equations:

$$t_e = 2(\varphi R_g + t_g) - t_i \tag{5.69}$$

$$t_i = \frac{\varphi \Delta \tau}{\rho_f c_f m} + t_e \tag{5.70}$$

where t_e is the outlet water temperature of the GHE, in °C; t_i is the inlet water temperature of the GHE, in °C; t_g is the ground temperature adjacent to boreholes, which can be obtained using Eqn (5.65), in °C; R_g is the thermal resistance between circulating water and ground adjacent to boreholes including convective thermal resistance between circulating water and pipe wall, pipe thermal resistance and backfilled material thermal resistance and thermal contact resistance between ground and pipe, in (m °C)/W; ρ_f is the water density, in kg/m^3; c_f is the specific heat of water, in J/(kg °C); m is the water flow rate, in m^3/s.

The Eqns (5.66)–(5.70) are used for calculating boundary conditions of the Eqn (5.65). The Eqn (5.65) is then used to calculate ground temperature distributions. Iterations need to be done because these equations are coupled.

5.4.8.2 Model Validation

The ground initial temperature is 17.6 °C, the thermal conductivity is 1.537 W/(m °C), the specific heat capacity is 2156 J/(kg °C) and the ground density is 1871 kg/m^3. These values were retrieved from field measurements of ground thermal properties in Nanjing, China [46]. A total of 25 boreholes with each borehole 50 m in depth are arrayed as shown in Figure 5.18. To investigate impacts of the heat balance, daily running mode and the space between boreholes, the scenarios listed in Table 5.16 were modelled.

Beier et al. [48] carried out an experiment to obtain reference data sets for a vertical borehole GHE. A sandbox of 18 m × 1.8 m × 1.8 m with a borehole at the centre was constructed for measurements. These experimental results were used for validation of the described model. Figure 5.19 shows the comparison between the measured and the predicted water temperatures into and out of the GHE [46]. The experimental data and model prediction results are in good agreement.

Table 5.16 Different Scenarios for Evaluate GCHP Performance

Scenario	Period	Running Time (h/day)	Cooling Load/ Heating Load (kW)	Space (m)	Number of Boreholes
1	Jun–Aug (Summer)	24	50/0	5	5 × 5
2	Jun–Aug (Summer) Nov–Jan (Winter)	24	50/ − 40	5	5 × 5
3	Jun–Aug (Summer) Nov–Jan (Winter)	24	40/ − 50	5	5 × 5
4	Nov–Jan (Winter)	24	0/ − 50	5	5 × 5
5	Jun–Aug (Summer)	8	50/0	5	5 × 5
6	Jun–Aug (Summer)	10	50/0	5	5 × 5
7	Jun–Aug (Summer)	12	50/0	5	5 × 5
8	Jun–Aug (Summer)	24	50/0	1	5 × 5
9	Jun–Aug (Summer)	24	50/0	3	5 × 5
10	Jun–Aug (Summer)	24	50/0	7	5 × 5
11	Jun–Aug (Summer)	24	50/0	5	7 × 7

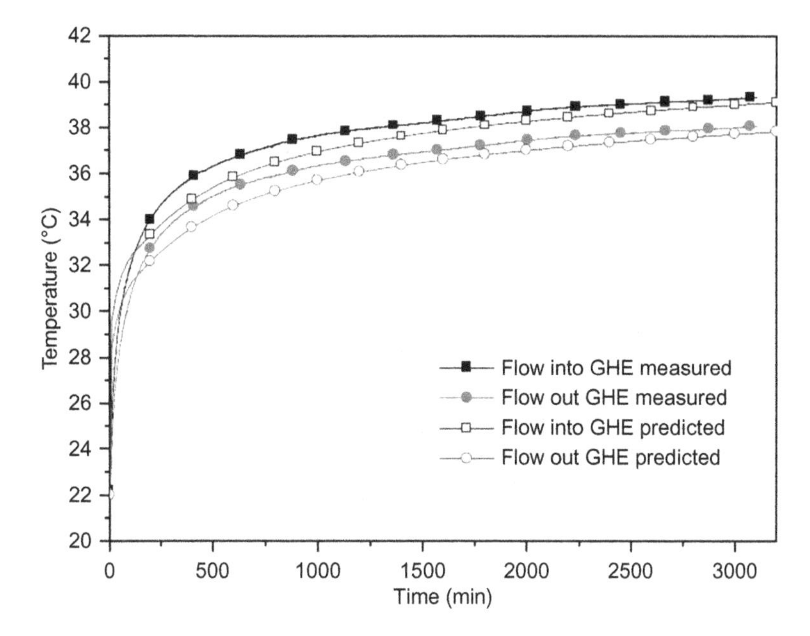

Figure 5.19 Comparison of the measured and the predicted GHE inlet and outlet water temperatures.

The temperature difference between the experiment and the prediction is within 0.2 °C at the end of the modelling time period.

5.4.8.3 The Impact of Seasonal Balance Between Heat Extraction and Heat Rejection

Heat or cold accumulation substantially influences the HP COP. Unbalanced heat injected into ground in the summer and extracted from ground in the winter may lead to heat or cold accumulation. Four different scenarios were simulated [46] to demonstrate the seasonal balance between heat extraction and heat rejection as shown in Table 5.16.

- Scenario 1 (50, 0 kW). The GCHP only runs in the summer (June–August) and the cooling load of the buildings is 50 kW. The heat injected into ground will change with the COP as shown in Eqn (5.66), which is larger than cooling load. If COP = 4, the total injected heat will be 62.5 kW, and the heat exchange between ground and borehole is 50 W/m.
- Scenario 2 (50, −40 kW). The GCHP runs in the summer with the cooling load of 50 kW and in the winter (November–January) with the heating load 40 kW. In the winter, heat extracted from ground is less than the heating load, as seen in Eqn (5.67). If COP = 4, the ratio of heat injection in the summer and heat extraction in the winter is 2.1.
- Scenario 3 (40, −50 kW). The GCHP runs in the summer with the cooling load of 40 kW and runs in the winter with the heating load of 50 kW. The ratio of heat injection in the summer and extraction in the winter is 1.33 when COP = 4.
- Scenario 4 (0, −50 kW). The GCHP only runs in the winter with the heating load of 50 kW.

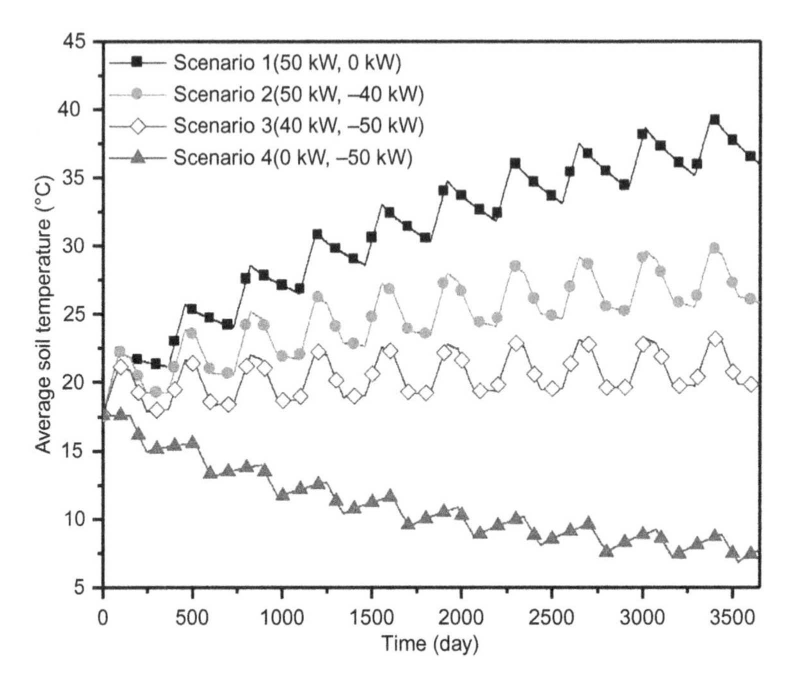

Figure 5.20 Variation of ground temperatures in 10 years with different heat flux ratios.

Figure 5.20 shows the results of the four simulation scenarios [46]. The heat accumulation or cold accumulation was detected when the building only had a cooling load or heating load. For a 10-year running period, the soil temperature for scenario 1 and scenario 4 increased to 39.1 °C and decreased to 6.9 °C, respectively. Based on our field observations, the GCHP system will not be able to perform properly when the soil temperature increases to 39.1 °C and decreases to 6.9 °C.

Scenario 1 and scenario 4 are for the GCHP running for 3 months during each year. Running for a longer period will make the heat or cold accumulation issue even more severe. For scenario 1, the average soil temperature increased rapidly in the cooling period when heat was injected into soil, and the temperature decreased slowly in the other period. When the soil temperature increased, the soil recovery speed increased as well due to the increment of the temperature difference between the soil around boreholes and soil further from the boreholes. The same phenomenon is observed in scenario 4. When the heat flux ratio was close to 1, as in scenario 3, the average soil temperature fluctuated with the initial soil temperature. There is also a gradual, but small increase in the annual average temperature for the first several years. It is because the HP starts to run from the summer, which means the heat will be firstly injected into the soil in the summer and then be extracted from the soil in the winter. This will lead to the gradual small increase in the annual average temperature.

In general, the COP value of scenario 1 and scenario 4 slightly decreased year by year. The COP values of scenario 3 were relatively constant. The COP values for scenario 2 in the summer decreased year by year while it increased year by year in the winter, which is because of the increase of the soil temperature.

These results indicate that seasonal balance between heat collection and rejection is very important for GCHP. The GCHP installed in buildings with both cooling and heating loads perform better compared with those in buildings with only cooling or heating load. Extending the length of boreholes and the space between boreholes may overcome this drawback when the heating and cooling loads are not balanced.

5.4.9 Vertical GHE Design Based on Hourly Load Simulations

Most design programs for GHE design use a superposition of annual, monthly and hourly pulses. Some design methods use monthly and hourly loads. Hourly load simulations are used mostly for simulation purposes, but not for design due to longer simulation times. Lamarche et al. [49] developed a new design approach based on hourly load data with the aid of an accelerated algorithm. This new approach does not use an aggregation of loads and simulates the real thermal response of the individual hourly loads.

5.4.9.1 Borehole Length Design

Kavanaugh and Rafferty [34] suggest the following relation for the borehole length for cooling loads:

$$L = \frac{q_a R_{ga} + (q_{lc} - W_c)(R_b + PLF_m R_{gm} + R_{gd} F_{sc})}{t_g - \frac{t_i + t_e}{2} - \Delta t_g} \tag{5.71}$$

where q_a is the annual average heat transfer to the ground; q_{lc} is the building cooling load; W_c is the power demand at the design cooling load; R_b is the borehole thermal resistance; PLF_m is the part-load factor during the design month; F_{sc} is the short-circuit heat loss due to heat transfer between the two different legs of the U-tube in the borehole; t_g is the ground temperature; t_i and t_e are the inlet–outlet fluid temperatures; and Δt_g is the so-called penalty temperature, representing the long term interference effect. The values of R_{ga}, R_{gm} and R_{gd} represent the effective ground thermal resistances for three thermal pulses, an annual pulse of 10 years, a monthly pulse of 1 month and a daily pulse of 6 h, respectively.

A similar expression is given for the heating loads, and the maximum length is chosen at the end. The borehole exit temperature is computed using the equation [49]:

$$t_e = t_g + \frac{q_i}{2mc_f} + \frac{q_i}{L} R_b + \frac{1}{2\pi\lambda} \sum_{i=1}^{N} (q_i - q_{i-1}) f\left(\frac{\tau_i - \tau_{i-1}}{\tau_s}, \frac{r_b}{L}\right) \tag{5.72}$$

where the length L is iterated until the exit temperature falls within the desired design value, and $\tau_s = L^2/9a$ is the steady-state time. The summation is performed on monthly periods except for the last few hours (hours 3 or 6), where hourly load is used.

Table 5.17 CPU-Time, in s

Simulation Time	No Aggregation	Lamarche	Yavuzturk	Bernier
1 month	16	0.02	16	1.7
6 months	554	0.10	135	7.4
1 year	2130	0.20	283	14
5 years	53,902	1.10	1470	71
10 years	222,830	1.90	3090	134

The scheme uses the concept of *f*-function developed by Eskilson [26] for the evaluation of the thermal response of the ground.

5.4.9.2 Design Algorithm

Basically, the borehole temperature is computed using a recurrence formula in a form similar to:

$$t_b(\text{Fo} + \Delta\text{Fo}) - t_g = \frac{1}{\lambda}\sum_{n=1}^{N} \underbrace{[e^{-z^2\Delta\text{Fo}}F_n(\text{Fo})(1 - e^{-z^2\Delta\text{Fo}})u_n(z)]\Delta z_n}_{F_n(\text{Fo}+\Delta\text{Fo})} \tag{5.73}$$

where $\text{Fo} = a\tau/r^2$ is the Fourier number and $F_n(\text{Fo})$ are recurrence coefficients computed from previous values as shown in Eqn (5.73).

The computational scheme has been developed based on the classical cylindrical heat source function proposed by Carslaw and Jaegers [23]. In that case, $u(z)$ has the following analytical expression [49]:

$$u(z) = \frac{2}{\pi^3}\frac{1}{z^3[J_1(z)^2 + Y_1(z)^2]} \tag{5.74}$$

The computational scheme is based on the numerical evaluation of the inverse Laplace transform:

$$u(z) = -2zL^{-1}(f/2\pi) \tag{5.75}$$

where f is the so-called *f*-function associated with the particular bore field. The tabular values calculated by Eskilson [26] can be used. An analytical solution was used for these functions [50].

To evaluate the numerical efficiency of this scheme, Table 5.17 compares the central processing unit (CPU) time to some aggregation schemes proposed in the past [25,51].

The total length of the GHE is iterated until the HP entrance temperatures (or the ground outlet temperatures) meet the constrained design values. The use of this efficient scheme allows us to evaluate more precisely the physical effects, such as short-time thermal behaviour, that will give more precise evaluation of the required borehole length.

The flow chart given in Figure 5.21 describes the design algorithm.

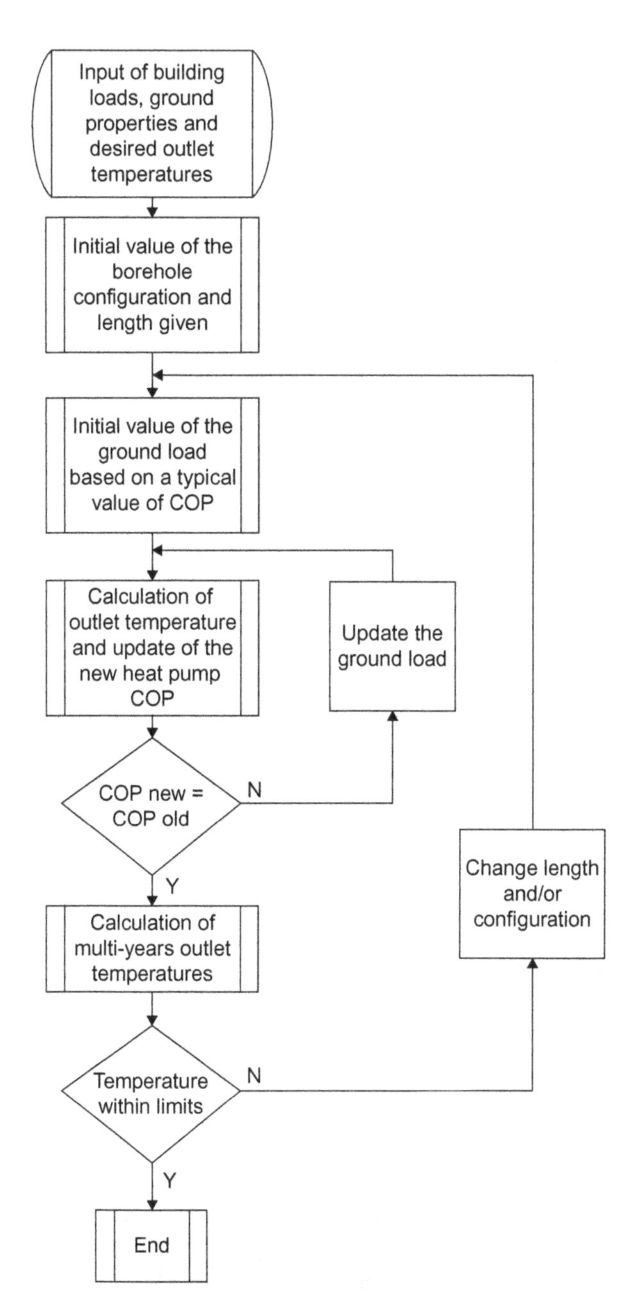

Figure 5.21 Design algorithm.

5.4.9.3 Results

The described design algorithm, referred to as the hourly load simulation design (HLSD), will be compared to two other computational models: TRNSYS and EED.

The first example is based on a residential application in a northern climate (Montreal, Canada) [49]. The building loads are typical of this type of climate, with a high heating demand in winter and moderate cooling during the summer.

The building loads calculations were performed using the TRNSYS simulation software. The computed annual loads are shown on Figure 5.22.

In the simulation, the load chosen was large enough to have more than one borehole. In this case, the effect of long-term thermal interference between the boreholes will be considered.

These loads were than used in another TRNSYS project that includes a vertical GHE system. The simulation was run over 10 years with a 0.1 h time step to ensure convergence. The total length of the GHE was modified until the exit ground temperature constraints were met.

Maximum entering water temperature (EWT) in the HP:

$$\max \mathrm{EWT} < t_g + [10, 15]^\circ\mathrm{C}$$

Minimum HP EWT:

$$\min \mathrm{EWT} > t_g - [5, 10]^\circ\mathrm{C}$$

The ground temperature t_g used was 8 °C. The maximum EWT in the HP was fixed at 13 °C, and the minimum EWT was fixed at 1 °C. The input data for the bore field are summarised in Table 5.18.

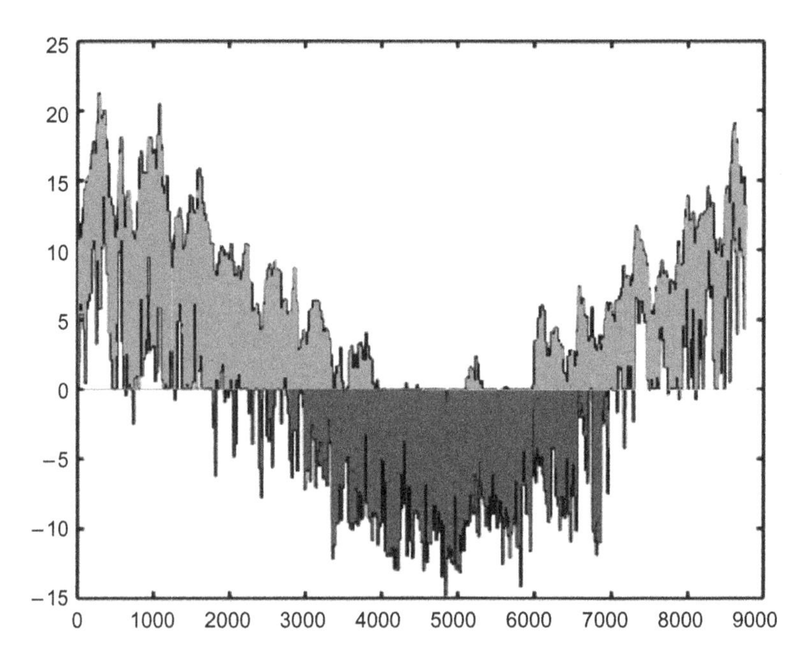

Figure 5.22 Annual building loads.

Table 5.18 Bore Field Configuration

Specification	Value
Bore diameter	0.102 m
Tube OD	0.034 m
Tube ID	0.014 m
Tube conductivity	0.4 W/(m K)
Shank spacing	0.05 m
Ground conductivity	1.3 W/(m K)
Grout conductivity	1.3 W/(m K)
Ground temperature	8 °C
Temperature gradient	0 °C/m
Configuration	2 × 2 rectangular; 5 m spacing
Number of years simulated	10 years

Table 5.19 Monthly Loads, in MWh

Month	Heating Demand	Cooling Demand
January	3.41	0
February	2.64	0
March	1.88	0
April	0.89	0.079
May	0.24	0.440
June	0	0.897
July	0	1.289
August	0	0.998
September	0.19	0.569
October	0.78	0.207
November	1.85	0
December	3.10	0

In the first test, the length calculation was compared to the software program EED, which uses monthly loads. These loads were found from the hourly loads given by TRNSYS. Table 5.19 gives the monthly loads, and Table 5.20 gives the peak load demand also needed in the design calculation. The total length computation results are given in Table 5.21.

The CPU time for the two design programs was very small, 10 s or less. For TRNSYS, each 10 year simulation run took approximately 5 min. To reach the desired output temperatures, four to five runs were typically performed until the constraints were met.

Table 5.20 Peak Demand, in kW

Month	Heating	Cooling
January	8.99	0
February	8.67	0
March	6.74	0
April	4.50	3.46
May	2.85	5.22
June	0	5.72
July	0	6.29
August	0	6.21
September	2.62	5.12
October	3.81	5.05
November	5.87	0
December	8.05	0

Table 5.21 GHE Length

Model	Length (m)	Relative Error (%)
TRNSYS	340	–
EED	358	5.3
HLSD	334	1.8

The main goal was to compare the ground model and to see the effect of using hourly loads versus monthly loads. For these reasons, all other parameters were fixed with those given by the TRNSYS simulation. For example, the same HP model was used in HLSD and TRNSYS. The same lookup table for the COP calculations was also used. The EED use a fixed COP value, so it was not possible to simulate the same HP. An average value for the heating COP was then given for these software programs. The EED uses the average fluid temperature as the input constraint instead of the output temperature. The value of $-0.55\ ^{\circ}C$ was chosen because it corresponded to the average fluid temperature calculated in TRNSYS.

Another very important aspect was the influence of the borehole resistance. This value has a very important influence on the calculation of the total heat exchanger length. To avoid large variations due to the calculation of this resistance, the value computed by TRNSYS was obtained and used in HLSD. In EED, it is possible to let the software compute this resistance or to provide it as an input.

5.4.10 Hybrid GCHP Systems

It is known that the GCHP systems can achieve better energy performance in specific locations where building heating and cooling loads are well balanced all the year round because of the long-term transient heat transfer in the GHEs. However, most

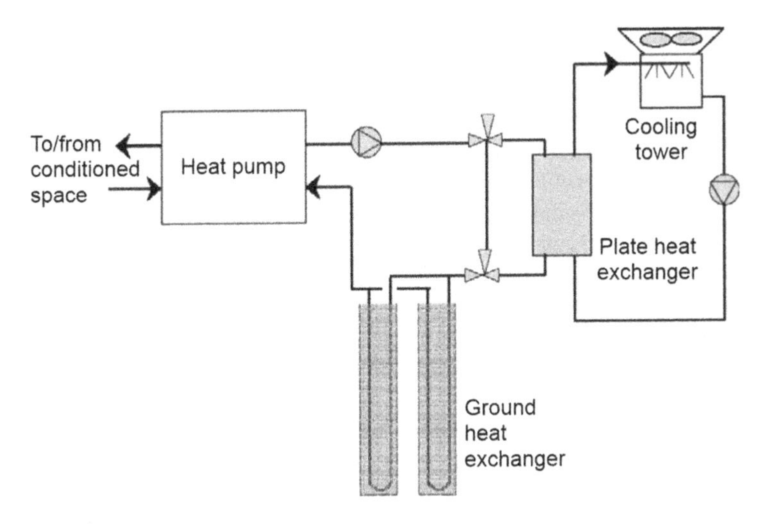

Figure 5.23 Schematic diagram of a HGCHP with cooling tower.

buildings in warm-climate or cold-climate areas have unbalanced loads, dominated by either cooling loads or heating loads.

An alternative to decrease the initial cost of the GCHP system and, at the same time, to improve the system performance is to employ a supplemental heat rejecter or heat absorber, which is called the hybrid GCHP (HGCHP) system [19].

5.4.10.1 HGCHP Systems with Supplemental Heat Rejecters

Figure 5.23 shows the operation principle of the HGCHP system with a cooling tower, where the cooling tower is connected in series with the GHE loop and is isolated from the building and ground loops with a plate heat exchanger.

The ASHRAE manual [6] discussed the advantages of the HGSHP applications for cooling-dominated buildings, considering initial costs and available surface are limitations. Kavanaugh and Rafferty [34] have discussed the possibility of the HGCHP system with a fluid cooler as a favourable alternative to lower the initial cost of the GHE installation. They recommended that the hybrid system be sized based on the peak building load at the design condition and the capacity of the cooler be calculated according to the difference between the GHE lengths required for cooling and heating loads.

Kavanaugh [52] has proposed a revised design method for sizing fluid coolers and cooling tower for hybrid system on the basis of the design procedure by ASHRAE [6] and Kavanaugh and Rafferty [34]. In addition to sizing the GHE and cooler, this revision also provides a method for balancing the heat flow into the ground on an annual basis.

Yavuzturk and Spitler [51] have investigated the advantages and disadvantages of various control strategies for the operation of a HGCHP system with a cooling tower under different climatic conditions. The investigated control strategies are broadly categorised into three groups: (i) at point control for the HP entering or exiting fluid temperatures to activate the cooling tower; (ii) differential temperature control to operate the cooling tower when the difference between the HP entering or exiting temperature and the ambient wet-bulb temperature is greater than a set value, and (iii) scheduled control to operate the cooling tower during the night to accomplish the cool storage in the ground and avoid a long-term temperature rise. The simulation results for a small building indicate that the hybrid application appears to have significant economic benefit compared with the conventional system.

A practical hourly simulation model of the HGCHP system with a cooling tower was developed with the aim of analysing and modelling the heat transfer process of its main components on an hour-by-hour basis by Man et al. [53]. Hourly operation data of the HGCHP system are calculated by the developed computer program based on the hourly simulation model. The impacts of four different control strategies on performances of two different HGCHP systems are compared in this simulation.

5.4.10.2 HGCHP Systems with Solar Collectors

In heating-dominated climates, the single GCHP system may cause a thermal heat depletion of the ground, which progressively decreases the HP's entering fluid temperature. As a result, the system performance becomes less efficient. Similar to the cases of cooling-dominated buildings, the use of a supplemental heat supply device, such as a solar thermal collector, can significantly reduce the GHE size and the borehole installation cost. Basically, the GHE is sized to meet the cooling load and the supplemental heater is sized to meet the excess heating load that is unmet by the GHE. Figure 5.24 shows the basic operating principle of the hybrid GCHP system with a solar collector.

The idea to couple a solar collector to the coil of pipes buried in the ground, by means of which solar energy can be stored in the ground, was first proposed by Penrod in 1956. Recently, a number of efforts have been made to investigate the performance and applications of the solar-assisted GCHP systems. Chiasson and Yavuzturk [54] presented a system simulation approach to assess the feasibility of the hybrid GCHP systems with solar thermal collectors in heating-dominated buildings. Yuehong et al. [55] conducted the experimental studies of a solar-ground HP system,

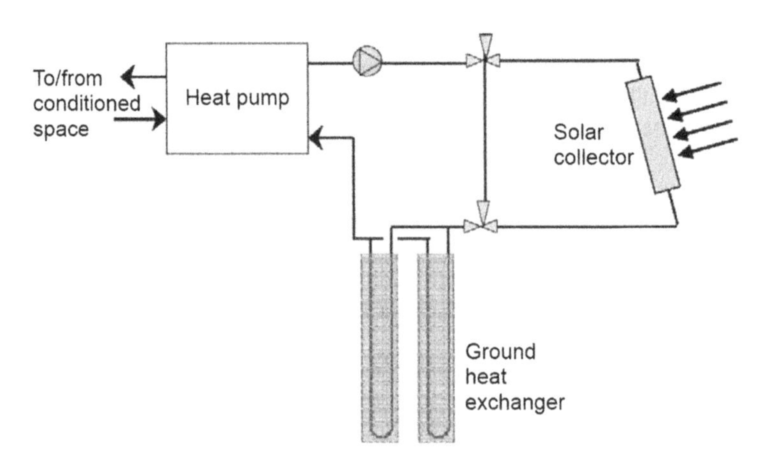

Figure 5.24 Schematic diagram of a HGCHP system with solar collector.

where the heating mode is alternated between a solar energy-source HP and a GSHP with a vertical double-spiral coil GHE. Ozgener and Hepbasli [56] experimentally investigated the performance characteristics of a solar-assisted GCHP system for greenhouse heating with a vertical GHE.

A solar-assisted GCHP heating system with latent heat energy storage tank (LHEST) was investigated by Zongwei et al. [57]. The hybrid heating system can implement eight different operation models according to the outdoor weather conditions by means of alternative heat source changes among the solar energy, ground heat and the LHEST. Finally, it is claimed that the LHEST can improve the solar fraction of the system, and thus the COP of the heating system can be increased.

5.5 ENVIRONMENTAL PERFORMANCES

The GSHPs work with the environment to provide clean, efficient and energy saving heating and cooling year round. GSHPs use less energy than alternative heating and cooling systems, helping to conserve natural resources. These are an important technology for reducing emissions of gases that harm the environment, such as carbon dioxide (CO_2), sulphur dioxide (SO_2) and nitrogen oxides (NO_x).

HPs driven by electricity from, for instance, hydropower or renewable energy reduce emissions more significantly than if the electricity is generated by coal, oil or natural gas power plants. The CO_2 emissions calculated for different primary energy sources [58] are summarised in Table 5.22.

The GSHPs utilise renewable or solar energy stored in the ground near the surfaces. The renewable component (66%) displaces the need for primary fuels, which, when burned, produce GHG emissions and contribute to global warming. An analysis was performed [59] to estimate the total equivalent warming impact of GSHPs compared to other heating and cooling systems in residential, commercial and institutional buildings. The modelling results show CO_2 emissions reductions from 15% to 77% through the application of GSHPs in both residential and commercial buildings.

The unique flexibility of GSHPs allows them to be used for residential and commercial buildings all across the United States, Canada and Europe. Regarding

Table 5.22 CO_2 Emissions for Different Primary Energy Sources

No.	System	Efficiency	CO_2 Emission per kWh of Fuel (kg CO_2/kWh)	CO_2 Emission per kWh of Useful Heat (kg CO_2/kWh)
1	Coal boiler	0.70	0.34	0.49
2	Gas-oil boiler	0.80	0.28	0.35
3	LPG boiler	0.80	0.25	0.31
4	Natural gas boiler	0.80	0.19	0.24
5	Air-to-air heat pump	2.50	0.47	0.19
6	Ground-to-water heat pump	3.20	0.47	0.15

CO_2 emissions, it could be seen that GSHPs do compete with condensing boilers in countries like Germany, United Kingdom and United States [60]. With increasing proportions of electricity generated from renewable sources, installing HPs in existing buildings becomes a more and more attractive option with respect to both primary energy demand and CO_2 emissions.

5.6 BETTER ENERGY EFFICIENCY WITH COMBINED HEATING AND COOLING BY HEAT PUMPS

The possibilities of HP solutions in combined cooling and heating systems have been unclear to a major portion of the designers of the A/C systems. Therefore, a survey was made to find out a proper dimensioning and disseminate the know-how. More general study was performed find out the influence of different factors.

A general study was carried out with a simple modelling tool [61]. The required heating and cooling capacities were calculated as the time-series using a simple dependence on outdoor temperature and solar radiation because the goal was to compare different systems and sizing of the HP. Variations of a heat source temperature of the HP are important for the annual COP. The presumed curves and the influence on COP are shown in Figure 5.25.

When the sizing factor (capacity) SF of HP increases the COP decreases (Figure 5.26) because a greater part of heating demand is produced under less favourable conditions, at lower heat source temperature. If the HP is dimensioned only for A/C cooling duty the SF is 40%.

Free cooling using the low temperature of the heat source is an effective way to decrease energy consumption of the compressor-based cooling. The temperature level of the heat source and the annual cooling demand profile determine how big a part can be covered by free cooling, as illustrated in Figure 5.27. Also, the temperature level of the cooling-water network has an essential influence: the higher the temperature, the bigger the part that can be produced by free cooling.

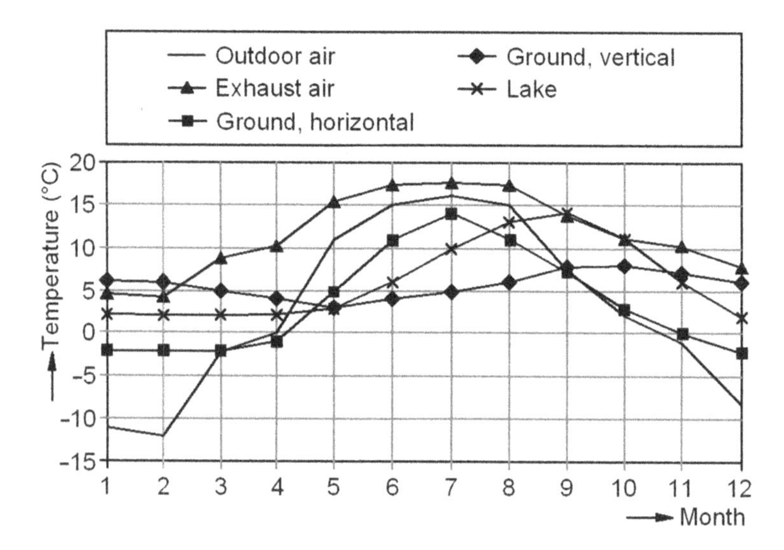

Figure 5.25 Temperature profiles of heat sources as used in modelling.

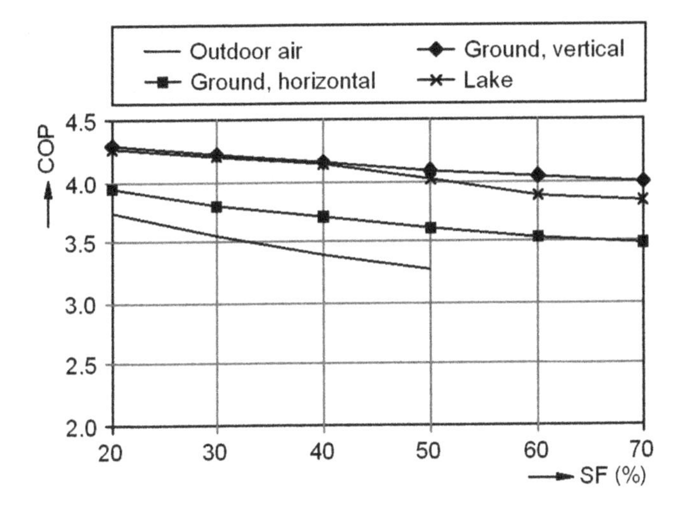

Figure 5.26 Annual COP as a function of sizing factor (SF).

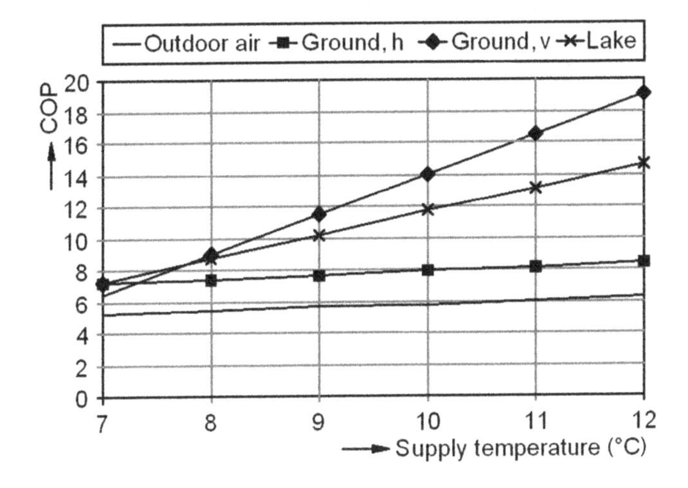

Figure 5.27 Influence of supply temperature of cooling water-network on COP for cooling.

Exhaust air as a heat source utilises heat after the normal heat recovery heat exchanger. When the efficiency on the heat exchanger is increased, the temperature before the evaporator falls and required capacity of the HP decreases. However, the electricity usage is almost constant, because COP decreases.

5.7 CONCLUSIONS

This chapter has provided a detailed theoretical study of GSHP technology, concentrating on GCHP systems and their applications. GSHPs are suitable for heating and cooling of buildings and could therefore play a significant role in reducing CO_2 emissions. During the past few decades, a large number of GSHP systems have been widely applied in various buildings around the world due to the systems' attractive advantages of high efficiency and environmental friendliness. GSHPs have increasingly been used for building heating and cooling with an annual growth rate of 10 – 15% in recent years.

The GWHPs have low costs, but with some limitations on the large water flow rate and the clogging of extraction wells with appreciable sediment quantities. The new GWHP system 'Geotherm', which has a COP of 4, removes these disadvantages through the use of a special heat exchanger.

Most typical heat transfer simulation models and ground TRT models currently available for vertical GHEs have been described in detail. Through the ground TRT, the length of a vertical GHE can be properly determined and supplementary costs (i.e. extra loops, boreholes, glycol, etc.) can be avoided. A one-dimensional transient GHE model was described to account for fluid and grout thermal capacities in BHE with the objective of predicting the outlet fluid temperature for varying inlet temperatures and flow rates. A novel model for calculating the ground temperature distribution and the COP of GCHP systems and a new design model for vertical GHEs were presented. In addition, the various hybrid GCHP systems for cooling- or heating-dominated buildings have been discussed.

Moreover, the application of existing GSHPs in already improved standard buildings with reduced heat demand and the development and market introduction of new high-temperature HPs are major task for the replacement of conventional heating systems with HPs in existing buildings.

REFERENCES

[1] Pahud D, Mattthey B. Comparison of the thermal performance of double U-pipe borehole heat exchanger measured *in situ*. Energy Build 2001;33(5):503–7.
[2] Bose JE, Smith MD, Spitler JD. Advances in ground source heat pump systems – an international overview. In: Proceedings of the 7th international conference on energy agency heat pump, Beijing, China; 2002. p. 313–24.
[3] Luo J, Rohn J, Bayer M, Priess A. Modeling and experiments on energy loss in horizontal connecting pipe of vertical ground source heat pump system. Appl Therm Eng 2013;60:55–64.
[4] Sarbu I, Sebarchievici C. General review of ground-source heat pump system for heating and cooling of buildings. Energy Build 2014;70(2):441–54.
[5] ASHRAE handbook, HVAC applications. Atlanta, GA: American Society of Heating, Refrigerating and Air–Conditioning Engineers; 2011.
[6] ASHRAE, Commercial/institutional ground-source heat pump engineering manual. Atlanta, GA: American Society of Heating, Refrigerating and Air-Conditioning Engineers; 1995.
[7] Rawlings RHD, Sykulski JR. Ground source heat pumps: a technology review. Build Serv Eng Res Technol 1999;20(3):119–29.
[8] Floridesa G, Kalogirou S. Ground heat exchanger – a review of systems, models and applications. Renewable Energy 2007;32(15):2461–78.
[9] Philappacopoulus AJ, Berndt ML. Influence of rebounding in ground heat exchangers used with geothermal heat pumps. Geothermic 2001;30(5):527–45.
[10] STULZ. Heat pumps technical data. Germany; 2006.
[11] Popiel C, Wojtkowiak J, Biernacka B. Measurements of temperature distribution in ground. Exp Therm Fluid Sci 2001;25:301–9.
[12] VIESSMANN. Heat pump systems – design guide. Romania; 2002.
[13] Tinti F. Geotermia per la climatizzazione. Palermo, Italy: Dario Flaccovio Editore; 2008.
[14] Sarbu I, Bura H. Thermal tests on borehole heat exchangers for ground-coupled heat pump systems. Int J Energy Environ 2011;5(3):385–93.
[15] Omer AM. Ground-source heat pumps systems and applications. Renewable Sustainable Energy Rev 2008;12(2):344–71.

[16] Bernier M. Closed-loop ground-coupled heat pump systems. ASHRAE J 2006;48(9):13–24.
[17] Hellström G. Ground heat storage: thermal analyses of duct storage systems [Doctoral thesis]. Sweden: Department of Mathematical Physics, University of Lund; 1991.
[18] OCHSNER. Heat pumps book. Linz, Austria: Eigenverlag; 2001.
[19] Yang H, Cui P, Fang Z. Vertical-borehole ground-couplet heat pumps: a review of models and systems. Appl Energy 2010;87:16–27.
[20] Ingersoll LR, Plass HJ. Theory of the ground pipe source for the heat pump. ASHVE Trans 1948;54:339–48.
[21] Ingersoll LR, Zobel OJ, Ingersoll AC. Heat conduction with engineering geological, and other applications. New York, NY: McGraw – Hill; 1954.
[22] Bose JE, Parker JD, McQuiston FC. Design/data manual for closed-loop ground-coupled heat pump systems. Oklahoma State University for ASHRAE; 1985.
[23] Carslaw HS, Jaeger JC. Conduction of heat in solids. Oxford, UK: Claremore Press; 1947.
[24] Deerman JD, Kavanaugh SP. Simulation of vertical U-tube ground coupled heat pump systems using the cylindrical heat source solution. ASHRAE Trans 1991;97(1):287–95.
[25] Bernier MA. Ground-coupled heat pump system simulation. In: ASHRAE winter meeting CD, technical and symposium papers. ASHRAE; 2001. p. 739–50.
[26] Eskilson P. Thermal analysis of heat extraction boreholes [Doctoral thesis]. Sweden: University of Lund; 1987.
[27] Zeng HY, Diao NR, Fang ZH. A finite line-source model for boreholes in geothermal heat exchangers. Heat Transfer Asian Res 2002;31(7):558–67.
[28] Muraya NK, O'Neal DL, Heffington WM. Thermal interference of adjacent legs in a vertical U – tube heat exchanger for a ground-coupled heat pump. ASHRAE Trans 1996;102(2):12–21.
[29] Rottmayer SP, Beckman WA, Mitchell JW. Simulation of a single vertical U – tube ground heat exchanger in an infinite medium. ASHRAE Trans 1997;103(2):651–9.
[30] Gu Y, O'Neal DL. Development of an equivalent diameter expression for vertical U – tubes used in ground-coupled heat pumps. ASHRAE Trans 1998;104:347–55.
[31] Zeng HY, Diao NR, Fang ZH. Efficiency of vertical geothermal heat exchangers in ground source heat pump systems. Int J Therm Sci 2003;12(1):77–81.
[32] Diao NR, Zeng HY, Fang ZH. Improvement in modeling of heat transfer in vertical ground heat exchangers. HVAC&R Res 2004;10(4):459–70.
[33] Cui P, Yang HX, Fang ZH. The simulation model and design optimization of ground source heat pump systems. HKIE Trans 2007;14(1):1–5.
[34] Kavanaugh SP, Rafferty K. Ground-source heat pumps, design of geothermal systems for commercial and institutional buildings. Atlanta, GA: ASHRAE; 1997.
[35] TRNSYS 17. A transient system simulation program user manual. USA: Solar Energy Laboratory, University of Wisconsin-Madison; 2012.
[36] Negut N. Operation book GEOTHERM PDC – first TRT in Romania, Bucharest; 2009 [in Romanian].
[37] Austin WA, Yavuzturk C, Spitler JD. Development of an *in-situ* system for measuring ground thermal properties. ASHRAE Trans 2000;106(1):365–79.
[38] Gehlin S. Thermal response test, *in-situ* measurements of thermal properties in hard rock [Licentiate thesis]. Sweden: Lulea University of Technology 1998;39:5–10.
[39] Bandyopadhyay C, Cosnold W, Mann M. Analytical and semi-analytical solutions for short-time transient response of ground heat exchangers. Energy Build 2008;40(10):1816–24.
[40] Beier RA. Equivalent time for interrupted tests on borehole heat exchangers. HVAC&R Res 2008;14(3):489–505.
[41] Martin CA, Kavanaugh SP. Ground thermal conductivity testing – controlled site analysis. ASHRAE Trans 2002;108(1):945–52.
[42] Agarwal RG. A new method to account for producing time effects when drawdown type curves are used to analyze pressure buildup and other test data. In: Proceedings of annual fall technical conference and exhibition of the society of petroleum engineers. Dallas Texas, September 21–24; 1980.
[43] Shirazi AS, Bernier M. Thermal capacity effects in borehole ground heat exchangers. Energy Build 2013;67:352–64.
[44] Cooper LY. Heating of a cylindrical cavity. Int J Heat Mass Transfer 1976;19:575–7.
[45] Bernier M, Pinel P, Labib R, Paillot R. A multiple load aggregation algorithm for annual hourly simulations of GCHP systems. HVAC&R Res 2004;10(4):471–88.
[46] Qian H, Wang Y. Modeling the interactions between the performance of ground source heat pumps and soil temperature variations. Energy Sustainable Dev 2014;23:115–21.

[47] Staffell I, Brett D, Brandon N, Hawkes A. A review of domestic heat pumps. Energy Environ Sci 2012;5(11):9291–306.
[48] Beier RA, Smith MD, Spitler JD. Reference data sets for vertical borehole ground heat exchanger models and thermal response test analysis. Geothermics 2011;40(1):79–85.
[49] Lamarche L, Beauchamp B. A fast algorithm for the simulation of GCHP systems. ASHRAE Trans 2007;113(1): DA-07-050
[50] Lamarche L, Beauchamp B. A new contribution to the finite line-source model for geothermal boreholes. Energy Build 2007;39:188–98.
[51] Yavuzturk C, Spitler JD. Comparative study of operating and control strategies for hybrid ground-source heat pump systems using a short time step simulation model. ASHRAE Trans 2000;106(2):192–209.
[52] Kavanaugh SP. A design method for hybrid ground-source heat pumps. ASHRAE Trans 1998;104(2):691–8.
[53] Man Yi, Yang H, Fang Z. Study on hybrid ground-coupled heat pump systems. Energy Build 2008;40(11):2028–36.
[54] Chiasson AD, Yavuzturk C. Assessment of the viability of hybrid geothermal heat pump systems with solar thermal collectors. ASHRAE Trans 2003;109:487–500.
[55] Bi Y, Guo T, Zhang L, Chen L. Solar and ground source heat pump system. Appl Energy 2004;78:231–45.
[56] Ozgener O, Hepbasli A. Performance analysis of a solar assisted ground-source heat pump system for greenhouse heating: an experimental study. Build Environ 2005;40(8):1040–50.
[57] Zongwei H, Maoyu Z, Fanhong K, Fang W, Li Z, Tian B. Numerical simulation of solar assisted ground-source heat pump heating system with latent heat energy storage in severely cold area. Appl Therm Eng 2008;28(11–12):1427–36.
[58] Laue HJ, Jakobs RM, Thiemann A. Energy efficiency and CO_2 reduction in the building stock – the role of heat pumps. Rehva J 2008;45(4):34–8.
[59] EPA. A short primer and environmental guidance for geothermal heat pumps. US Environmental Protection Agency; 1997.
[60] Huchtemann K, Muller D. Evaluation of a field test with retrofit heat pumps. Build Environ 2012;53:100–6.
[61] Aittomäki A. Better energy efficiency with combined heating and cooling by heat pumps. Rehva J 2009;46(3):29–31.

CHAPTER 6

Heat Pump Heating and Cooling Systems

6.1 GENERALITIES

The buildings sector is the largest user of energy and source of CO_2 emissions in the EU, and buildings are responsible for more than 40% of the EU's total energy use and CO_2 emissions.

Of the total energy consumption of a building, approximately 54% goes to heating. To cover this energy demand, great quantities of fossil fuel are burned, which generates considerable carbon dioxide (CO_2) emissions [1].

There are more than 150 million dwellings in Europe. Approximately 30% were built before 1940, approximately 45% between 1950 and 1980 and only 25% after 1980. Retrofitting is a means of rectifying existing building deficiencies by improving the standards and the thermal insulation of buildings and/or replacing old space conditioning systems with energy-efficient and environmentally-sound heating and cooling systems [2,3]. Furthermore, EU member states must stimulate the transformation of existing buildings undergoing renovation into nearly zero-energy buildings (nZEBs). Conversion of heating and cooling systems to ground source heat pumps and air-to-water heat pumps (HPs) is a well-proven measure for approaching nZEB requirements.

For the first time, the Renewable Energy Directive 2009/28/EC of the European Parliament recognised aerothermal, geothermal and hydrothermal energy as renewable energy sources. This directive opens up a major opportunity for further use of HPs for the heating and cooling of new and existing buildings [4].

The integration of renewable energy in nZEBs by 2020 is taken into consideration in the calculation of the primary energy factors. The primary energy factors will be lowered over time as renewable energy will comprise a larger proportion of the energy mix (Table 6.1).

Local, collective renewable energy installations such as wind turbines, shared solar heating systems, solar photovoltaic arrays or geothermal systems are included in the calculations if the building owner owns a share of the installation.

GCHP systems are a type of renewable energy technology that has been increasingly used in the past decade across Europe to provide air-conditioning (A/C) and domestic hot water (DHW) for buildings [5–7]. These systems can achieve higher energy efficiency compared to air-source HP systems because the soil can provide a lower temperature for cooling and a higher temperature for heating than air [8].

A GCHP system consists of a conventional HP coupled with a ground heat exchanger (GHE) where water or a water-antifreeze mixture exchanges heat with the ground. A GHE may be a simple pipe system buried in the ground; it may also comprise a horizontal collector or, more commonly, a borehole heat exchanger (BHE) drilled to a depth of between 20 and 300 m with a diameter of 100–200 mm.

Ground-Source Heat Pumps. DOI: http://dx.doi.org/10.1016/B978-0-12-804220-5.00006-0

Table 6.1 Primary Energy Conversion Factors

Primary/Useful Energy	2006	Low-Energy 2015	Building Class 2020
District heating	1.0	0.8	0.6
Fossil fuels	1.0	1.0	1.0
Bio fuels	1.0	1.0	1.0
Electricity	2.5	2.5	1.8

The widespread distribution of HPs as single generators in heating systems has mainly been in new, rather isolated buildings, that have limited unit loads. This has enabled the use of low-temperature terminal units, such as fan coil units and, in particular, radiant systems. After the introduction of plastic piping, the application of water-based radiant heating and cooling with pipes embedded in room surfaces (i.e. floors, walls and ceilings), has significantly increased worldwide. Earlier applications of radiant heating systems were mainly for residential buildings because of the comfort and free use of floor space without any obstruction from installations. For similar reasons, as well as possible peak load reduction and energy savings, radiant systems are widely applied in commercial and industrial buildings. Due to the large surfaces needed for heat transfer, the systems work with low water temperatures for heating and high water temperatures for cooling. However, to extend the use of these types of generators and to benefit from their energy efficiency to reach the targets of 20-20-20 (20% increase in energy efficiency, 20% reduction of CO_2 emissions, and 20% renewable by 2020), working with radiators, which were the most commonly used terminal units in heating systems in the past, is necessary.

There are tens of thousands of buildings to restore in Europe, the majority of which are residential. The energy challenge of the future will be in renovating existing buildings, and whoever proposes system-engineering technologies that can be installed with minimal interventions will be immensely successful. Therefore, if HP technology is promoted, it must be designed to also work with radiators.

This chapter compares different heating systems in terms of energy consumption, thermal comfort and environmental impact. A comparative economic analysis of heating solutions for a building is performed from a case study, and the energy and economic advantages of building heating solutions with a water-to-water HP are reported. The energy, economic and environmental performances of a closed-loop GCHP system is also analysed. In addition, the main performance parameters (energy efficiency and CO_2 emissions) of radiators and radiant floor heating systems connected to GCHPs are compared. These performances were obtained from site measurements in an office room. Furthermore, the thermal comfort for these systems is compared using the ASHRAE Thermal Comfort program [9] and a mathematical model for numerical modelling of the thermal emission at radiant floors is developed and experimentally validated. Additionally, two numerical simulation models of useful thermal energy and the system coefficient of performance (COP_{sys}) in heating mode are developed using the TRNSYS software. The simulations obtained from the TRNSYS software are analysed and compared to experimental measurements. Finally, important information for control of HP heating and cooling systems is included.

6.2 RADIATOR HEATING SYSTEM

A hot-water radiator heating system is a type of central heating. In this system, heat is generated in a boiler. For the generation of the heat, a natural gas boiler is used where the chemical energy of natural gas is transferred into the heat. Then, the heat is distributed by hot water (heat carrier) to the radiators. The radiators heat the rooms. The radiators are installed in each heated room of the house. The hot water is circulated by a water circulation pump, which operates continuously. If the valves stop, then the hot water flows through a bypass pipe. The radiators, as rule of thumb, are located next to the cold surfaces of the envelope. They significantly influence thermal comfort. The radiators release the highest amount of heat to the heated room by convection and one part by heat radiation.

The convective heat transfer will lead to a lower relative humidity of the air, and, at high radiator surface temperature, dust particles can be burned, leading to lower indoor air quality. Thus, emitters should be implemented with a radiation factor as high as possible in the case of high-temperature water supplies. The values of the radiation factor are presented in Table 6.2 for typical hot water radiators [10].

The highlights of the convective thermal field achieved with radiators, illustrated in Figure 6.1a, are as follows:

1. a warm-air jet (1) that is raised from the area of the heater to the upper part of the room, as a result of the gravitational forces
2. a warm-air jet (2), developed at the surface of the ceiling, also as a result of gravitational forces
3. a rotational area, for air circulation, which is more active on the vertical (3a) in the area of the warmth source and less active (3b) under the ceiling air jet
4. a steady area (4) in the middle of the room
5. induction currents (5) at the floor level resulting from air cooling by vertical walls.

Computer simulation studies using TRNSYS software indicate that according to the heat flux emitted by the heater, as well as to its temperature level, it is possible to establish

Table 6.2 Radiation Factor of Usual Radiators

Radiator Type	Heat Transferred by Radiation		
	Room-Wards	Wall-Wards	Total
Steel column radiator	0.28	0.10	0.38
Cast-iron column radiator	0.26	0.10	0.36
Panel radiator			
1/0[a]	0.38	0.18	0.56
1/1	0.25	0.11	0.36
2/0	0.23	0.10	0.33
2/1	0.20	0.08	0.28
2/2	0.17	0.07	0.24
3/3	0.14	0.04	0.18

[a] *The first number represents the number of panels and the second is the number of convective elements.*

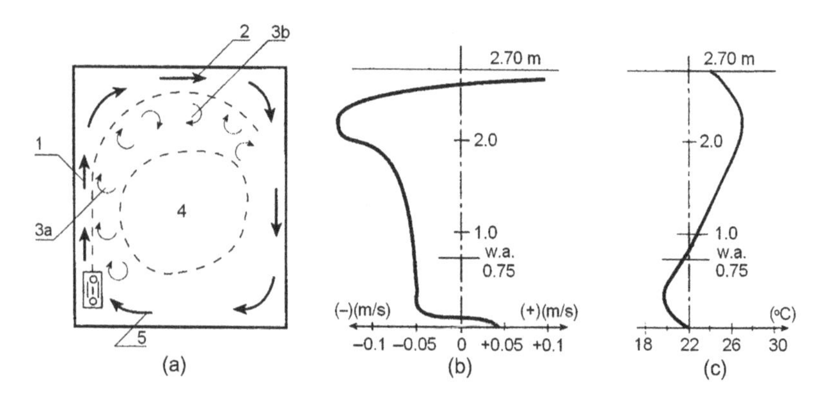

Figure 6.1 Convective field for the static heaters in living rooms. (a) Convective currents; (b) Velocity profile; (c) Temperature profile; w.a., Work area.

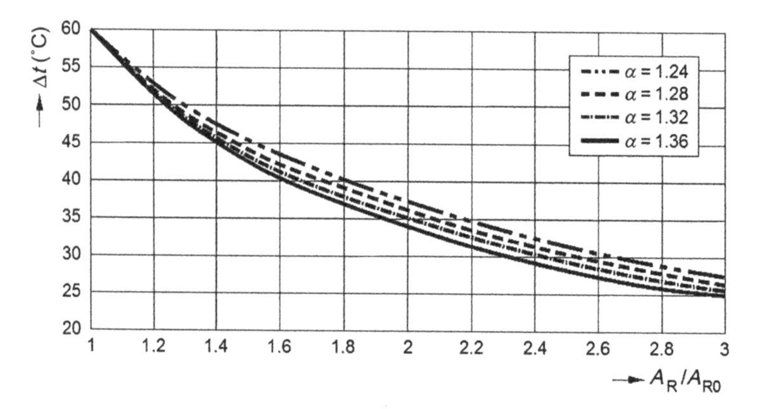

Figure 6.2 Variation of radiator surface.

a velocity profile (Figure 6.1b) related to the temperature profile (Figure 6.1c) in rooms with a certain given geometry. The heat flux was 2163 W per 1 m radiator length and inlet and outlet temperature to the radiator was 90 and 70 °C, respectively. Thus, for a living room (with height $h = 2.70$ m), the air current velocity to the middle of the room is practically constant, increasing in the active circulation area and afterwards, at the ceiling, it starts to decrease again. The air temperature increases substantially enough, from 20–22 °C in the working area, to 24–26 °C under the ceiling.

In regards to the heat flux values yielded by the heaters, *in situ* measurements [11] have indicated that the heat flux yielded by convection has the highest value. The heat flux yielded by radiation to the convectors is lower, for radiators it is under 50%, and for radiators with metal fins it is 10–25%.

The high temperature of the hot water can lead to a lower thermal comfort level because of the asymmetric radiation [12,13].

Figure 6.2 shows the variation of a radiator surface depending on the logarithmic mean temperature difference Δt for different values of the α radiator exponent. An increase of the radiator surface (A_R/A_{R0}) can be observed while the values of the α exponent decrease for the same temperature difference Δt. The necessary radiator surface A_R will increase for heating systems with supply/return water temperatures lower than 90/70 °C.

To ensure ever-changing heat demand in a room, qualitative, quantitative or mixed control systems are used. By qualitative control the controlled parameter is the supply temperature and the flow rate is constant during the operation time. By quantitative control the controlled parameter is the flow rate, the supply temperature remaining constant throughout the whole operation period.

6.3 RADIANT HEATING AND COOLING SYSTEMS

6.3.1 Description of the Systems

In low-energy buildings, the low-temperature heating system usually works with a supply water temperature below 45 °C [14]. Embedded radiant systems are used in all types of buildings.

Radiant heating application is classified as panel heating if the panel surface temperature is below 150 °C [15]. In thermal radiation, heat is transferred by electromagnetic waves that travel in straight lines and can be reflected. The water temperatures are operated at very close to room temperature and, depending on the position of the piping, the system can take advantage of the thermal storage capacity of the building structure.

The available types of embedded hydronic radiant systems [16] are usually insulated from the main building structure (i.e. floor, wall, ceiling), and the actual operation mode (heating/cooling) of the systems depends on the heat transfer between the water and the space.

Panel heating provides a comfortable environment by controlling surface temperatures and minimising air motion within a space. A radiant system is a sensible heating system that provides more than 50% of the total heat flux by thermal radiation. The controlled temperature surfaces may be in the floor, walls or ceiling, with the temperature maintained by circulation of water or air.

The radiant heat transfer is, in all cases, 5.5 W/(m^2 K). The convective heat transfer then varies between 0.5 and 5.5 W/(m^2 K), depending on the surface type and on heating or cooling mode. This shows that the radiant heat transfer varies between 50% and 90% of the total heat transfer [17].

The low-temperature radiant systems are very complex because they involve different mechanisms of heat transfer: heat conduction through the walls, heat convection between the heating panel and the indoor air, heat radiation between the heating panel and the surrounding areas and the heat conduction between the floor and the ground. The main goal of low-temperature radiant systems is to provide adequate thermal comfort at significantly lower temperatures.

Radiant panel heating is characterised by the fact that heating is associated with a yielding of heat with low temperature because of physiological reasons. Thus, at the radiant floor panels, the temperature must not exceed 29 °C, and at the radiant ceiling panels, the temperature will not exceed 35–40 °C, depending on the position of the occupier (in feet) and the occupier distance to the panels, in accordance with thermal comfort criteria established by ISO Standard 7730 [18]. For cooling, the

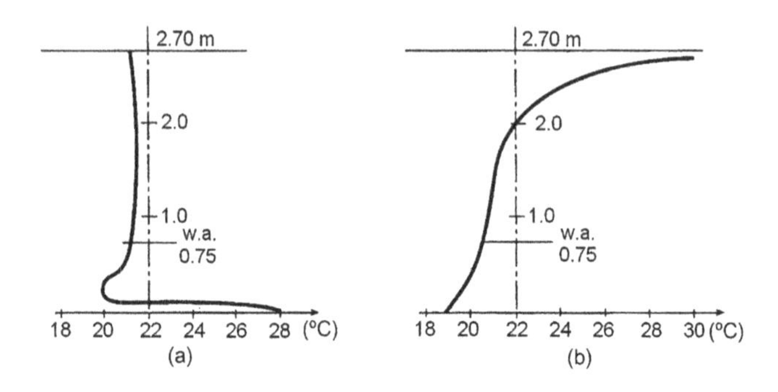

Figure 6.3 Temperature profile for low-temperature radiant heating. (a) Floor heating; (b) Ceiling heating.

minimum floor temperature is 19 °C. A vertical air temperature difference between head and feet of less than 3 °C is recommended. The vertical profile of the air temperature for two types of radiant heating panels is illustrated in Figure 6.3. The radiant part is lower (70%) at the floor heating than the ceiling in terms of heating (85%) because thermal convection is developed more in the case of floor heating panels [19].

The higher mean radiant temperature in a radiantly heated space means that the air temperature can be kept lower than in a convectively heated space. This has the advantage that the relative humidity in winter may be a little higher.

The heat transfer between the water and surface is different for each system configuration. Therefore, the estimation of heating/cooling capacity of systems is very important for the proper system design. Two calculation methods included in ISO 11855 are simplified calculation methods depending on the type of system, and finite element method or finite difference method.

The simplified calculation methods are specific for the given system types within boundary conditions. Based on the calculated average surface temperature at given heat carrier temperature and operative temperature in the space, it is possible to determine the steady-state heating and cooling capacity. Thus, the heating/cooling capacity of the systems is [20]:

- floor heating and ceiling cooling:

$$q = 8.92\left|t_o - t_{S,m}\right|^{1.1} \tag{6.1}$$

- wall heating and wall cooling:

$$q = 8\left|t_o - t_{S,m}\right| \tag{6.2}$$

- ceiling heating:

$$q = 6\left|t_o - t_{S,m}\right| \tag{6.3}$$

- floor cooling:

$$q = 7\left|t_o - t_{S,m}\right| \tag{6.4}$$

where: q is the heating/cooling capacity, in W/m^2; t_o is the operative (comfort) temperature in the space; $t_{S,m}$ is the average surface temperature.

The floor has the capacity of up to 100 W/m^2 for heating and 40 W/m^2 for sensible cooling. The ceiling has the capacity of up to 100 W/m^2 for sensible cooling and 40–50 W/m^2 for heating.

Control of the heating and cooling system needs to be able to maintain the indoor air temperatures within the comfort range under the varying internal loads and external climates. To maintain a stable thermal environment, the control system needs to maintain the balance between the heat gain/loss of the building and the supplied energy from the system.

6.3.2 Numerical Modelling of Thermal Emission at Radiant Floor

Floor heating construction corresponds with the mostly used system that is plastic tubing (PEX) embedded in the concrete slab.

6.3.2.1 Formulation of Mathematical Model

For heat exchange modelling between the radiant floor and environment, the 'virtual tube' method is used [21]. This allows the calculation of temperature at a point P around a tube T of radius ρ placed at a distance b from the surface S, maintained at a constant temperature of 0 °C.

The temperature at point P is given by the following equation:

$$t_P = \frac{t_{hc}}{\ln(\rho/2b)} \ln \frac{r}{r'} \tag{6.5}$$

where t_{hc} is the mean temperature of the heat carrier; r is the distance from point P to the centre of the tube cross section; r' is the distance from point P to the centre of the virtual tube cross section.

The virtual tube method was adapted to model thermal emission of radiant floors, considering that the distance b is the equivalent of the sum between the thermal resistances of floor layers above the tube (R) and the superficial heat transfer resistance (R_i). In this case, the surface S of constant temperature is represented by the environment (Figure 6.4).

Using Eqn (6.5) to calculate the temperature at point P located on the floor surface thus becomes the following equation:

$$t_P = t_i + \frac{t_{hc} - t_i}{\ln \frac{\rho}{2b}} \ln \frac{r}{r'} \tag{6.6}$$

in which

$$r = \sqrt{R^2 + x^2}; \quad r' = \sqrt{(R+2R_i)^2 + x^2} \tag{6.7}$$

$$R = \sum_{j=1}^{N} \frac{\delta_j}{\lambda_j}; \quad R_i = \frac{1}{\alpha_i} \tag{6.8}$$

where t_i is the indoor air temperature; R is the thermal diffusion resistance of compound floor layers above the tube, in m^2 K/W; R_i is the superficial heat transfer resistance at the internal surface, in m^2 K/W; δ_j and λ_j are the thickness, in m, and thermal conductivity of layer j, in W/(m K), respectively; α_i is the superficial heat transfer coefficient of the floor surface, in W/(m^2 K).

Therefore, the modelling allows for the determination of floor surface temperature at any point on its surface for different floor structures. In addition, the temperature at any point within the floor can be calculated. Thermal emission is

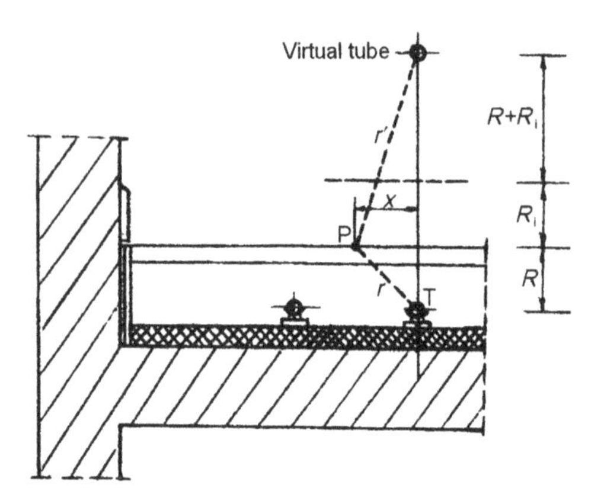

Figure 6.4 Schematics of radiant floor.

proportional to the difference between the surface temperature and indoor air temperature. However, the surface temperature is not uniform. This parameter has a nonlinear variation in relation to the distance from the vertical of tube section, according to Eqns (6.5) and (6.6).

The mean floor surface temperature is given by the following:

$$t_{f,m} = \frac{1}{s/2} \int_0^{s/2} \left[t_i + \frac{t_{hc} - t_i}{\ln \frac{\rho}{2(R+R_i)}} \ln \frac{r}{r'} \right] dx \tag{6.9}$$

Integrating Eqn (6.8) obtains the following:

$$t_{f,m} = \left[t_i + \frac{t_{hc} - t_i}{\ln \frac{\rho}{2(R+R_i)}} \right] \frac{A - B}{s} \tag{6.10}$$

in which

$$A = \left[\frac{s}{2} \ln \left(\frac{s^2}{4} + R^2 \right) \right] - \left[s - \left(2R \operatorname{arctg} \frac{s}{2R} \right) \right] \tag{6.11}$$

$$B = \left\{ \frac{s}{2} \ln \left[\frac{s^2}{4} + (R+2R_i)^2 \right] \right\} \left\{ s - \left[2(R+2R_i) \operatorname{arctg} \frac{s}{2(R+2R_i)} \right] \right\} \tag{6.12}$$

where s is the arrangement step of the radiant floor tubes, and A, B are notations.

6.3.2.2 Superficial Heat Transfer

Radiative heat transfer between the floor and the room walls is calculated with the following formula:

$$q_r = \sigma \varepsilon_1 \varepsilon_2 (T_1^4 - T_2^4) \tag{6.13}$$

in which q_r is the radiant flux, in W/m^2; $\sigma = 5.67 \times 10^{-8}$ W/(m^2 K^4) is the Stefan–Boltzmann constant; ε_1 and ε_2 are the thermal emittance of the floor surface and room walls, respectively (dimensionless); T_1 is the absolute temperature of radiant floor surface, in K; T_2 is the weighted average absolute temperature of all room walls, in K [15].

The radiative heat transfer coefficient α_r can be calculated by [22]:

$$\alpha_r = \frac{q_r}{T_1 - T_2} \tag{6.14}$$

Convective heat transfer is determined with a criteria group expressed by dimensionless numbers:

$$Nu = \frac{\alpha_c L}{\lambda} = C(Gr \cdot Pr)^n \tag{6.15}$$

$$Gr = \frac{g\beta L^3 \Delta t_{f-a}}{\nu^2} \tag{6.16}$$

$$Pr = \frac{\nu}{a} \tag{6.17}$$

in which Nu, Gr and Pr are Nusselt, Grashof and Prandtl numbers, respectively; α_c is the convective heat transfer coefficient, in W/(m^2 K); L is the characteristic dimension of the element surface, in m; λ is the thermal conductivity of air, in W/(m K); g is the gravitational acceleration, in m/s^2; C and n are the parameters depending on the $Gr \cdot Pr$ product; β is the volumetric expansion coefficient of air, in K^{-1}; Δt_{f-a} is the temperature difference between the floor surface and air, in K; ν is the kinematic viscosity of air, in m^2/s; a is the thermal diffusivity of air, in m^2/s.

6.3.2.3 Thermal Emission at Floor Surface

Thermal emission of radiant floor is computed by taking into account the weighted mean temperature of walls and indoor air temperature.

The operative (comfort) temperature t_o may be defined as the average of the mean radiant temperature t_r and indoor air temperature t_i weighted by their respective heat transfer coefficients [23]:

$$t_o = \frac{\alpha_r t_2 + \alpha_c t_i}{\alpha_r + \alpha_c} \tag{6.18}$$

in which α_r and α_c are the radiative and the convective heat transfer coefficient between body and environment, in W/(m^2 K), respectively. The mean radiant temperature is approximated with t_2.

The mean superficial heat flux q is given by the product of the total heat transfer coefficient $(\alpha_r + \alpha_c)$ and the difference between mean floor surface temperature $t_{f,m}$ and operative temperature t_o:

$$q = (\alpha_r + \alpha_c)(t_{f,m} - t_o) \tag{6.19}$$

The analytical model described above allows the determination of the maximum heat carrier temperature for any type of radiant floor structure. Exceeding this temperature leads to temperature values at the intersection point between the floor surface and the vertical of tube section beyond the maximum allowed 29 °C.

6.3.2.4 Computation Example

The radiant floor under consideration is used for an office room with geometrical dimensions of 6.7 m × 3.3 m × 3.45 m (Figure 6.5) located in Timisoara, Romania. The latitude and longitude of this city are 45°47′ N and 21°17′ E, respectively. The radiant floor has embedded plastic tubes (netlike polyethylene) with a diameter of 20 mm × 2 mm, arranged with a step $s = 20$ cm, through which circulates the heat

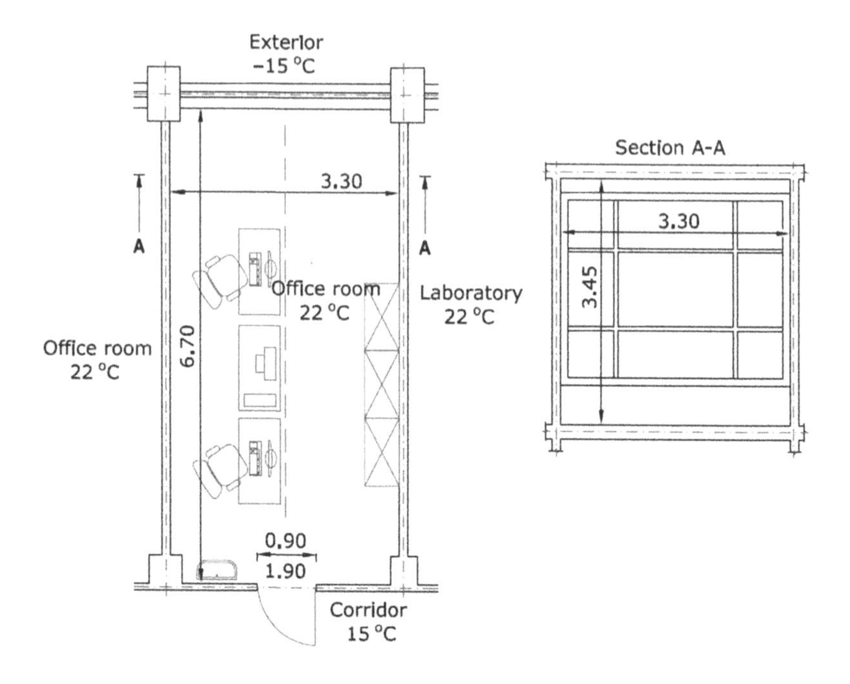

Figure 6.5 Heated office room.

Table 6.3 Numerical Computation Results

R (m^2 K/W)	R_i (m^2 K/W)	b (m)	t_P (°C)	$t_{f,m}$ (°C)	q_r (W/m^2)	α_r (W/m^2K)	α_c (W/m^2K)	t_o (°C)	q (W/m^2)
0.065	0.125	0.19	27.6	29.0	23.2	4.64	5.23	22.9	61.9

carrier of 42/36 °C. Above the tubes, there is a concrete layer with thickness $\delta_1 = 5$ cm and a wood layer with thickness $\delta_2 = 8$ mm. The indoor air temperature is 22 °C, and the weighted average wall temperature measured by thermograph is 24 °C. The thermal emittances are $\varepsilon_1 = 0.9$ and $\varepsilon_2 = 0.85$.

The floor temperature $t_{f,m}$ at a point situated in the centre of the distance between two consecutive tubes is determined, and finally, the operative temperature t_o and mean superficial heat flux q results are obtained. The numerical results obtained applying a previously developed computation model are summarised in Table 6.3.

6.3.2.5 Validation of Mathematical Model

Some statistical methods, such as the root-mean square (RMS), the coefficient of variation (c_v), the coefficient of multiple determinations (R^2) and percentage difference may be used to compare simulated (computed) and actual values for model validation.

The simulation error can be estimated by the RMS defined as [24]:

$$\text{RMS} = \sqrt{\frac{\sum_{i=1}^{n} (y_{\text{sim},i} - y_{\text{mea},i})^2}{n}} \tag{6.20}$$

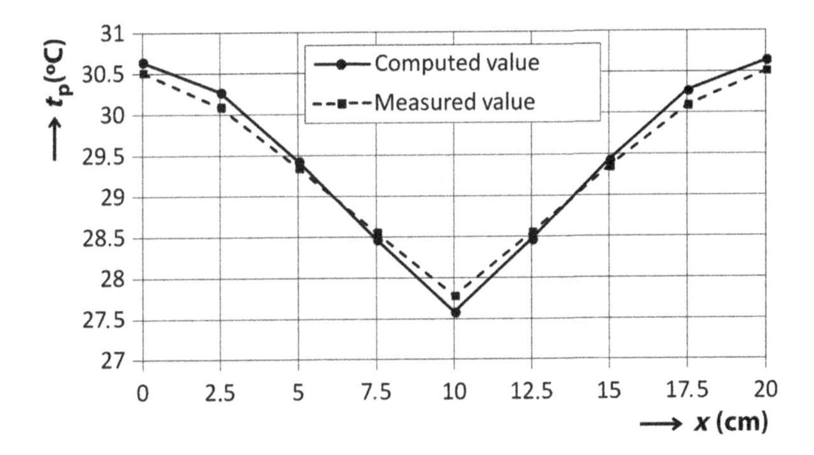

Figure 6.6 Graphic of computational and experimental results.

In addition, the coefficient of variation c_v, in percent and the coefficient of multiple determinations R^2 are defined as follows [24]:

$$c_v = \frac{\text{RMS}}{|\overline{y}_{\text{mea},i}|}100 \tag{6.21}$$

$$R^2 = 1 - \frac{\sum_{i=1}^{n} (y_{\text{sim},i} - y_{\text{mea},i})^2}{\sum_{i=1}^{n} y_{\text{mea},i}^2} \tag{6.22}$$

where n is the number of measured data in the independent data set; $y_{\text{mea},i}$ is the measured value of one data point i; $y_{\text{sim},i}$ indicates the simulated value; $\overline{y}_{\text{mea},i}$is the mean value of all measured data points.

The floor surface temperature values between two consecutive tubes, computed with the proposed model, are compared with measured values on a radiant floor heating system with the structure from the previous calculation example.

The temperatures measured at every point with a TESTO 350 instrument are represented by the average of three measurements. The maximum deviation between the extreme measured values was 0.15 °C. The obtained results are plotted in Figure 6.6, and show a good agreement between the experimental temperature measurements and the computed values. Statistical values such as RMS, c_v and R^2 are 0.12536, 0.00486 and 0.99998, respectively, which can be considered as very satisfactory. Thus, the computational model was validated by the experimental data.

6.4 TERMINAL UNIT SUPPLY TEMPERATURE

A radiant floor system requires a water input temperature between 35 and 40 °C. A radiator system built in the 1970s was designed with input temperatures higher than 70 °C. The question is, 'By how much can be the supply temperature of the radiators lowered whilst keeping the same size existing terminal units?' Both increasing the building performance and introducing an HP system results in a reduction in the requested power and permits a reduction of the water temperature sent to the plant.

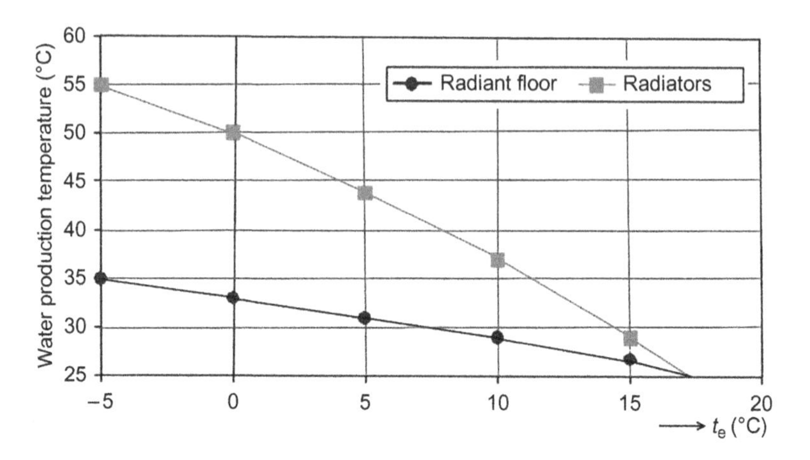

Figure 6.7 Water supply production temperature depending on outdoor air temperature.

The water input temperature in terminal units with both a radiant floor system and a radiator system can reduce the thermal load according to the change in outdoor air temperature t_e, as shown in Figure 6.7.

6.5 HEATING SYSTEM SELECTION

This book describes hydronic heat distribution systems (heating and cooling systems with water). Open systems are open to the atmosphere in at least one location. Closed systems, on the other hand, are not open to the atmosphere, except possibly at an expansion/compression tank. Both hot-water and chilled-water systems have common components that serve similar purposes. The components that they have in common include: piping, pumps, air separators, expansion tanks, fill accessories, valves and accessories.

As a rule, the lower the heating system temperature, the higher the COP of the HP and the lower the heating costs. To achieve this, the largest possible surface area for heat transfer must be selected. The ideal systems for this criterion are low-temperature radiant floor and wall heating systems (i.e. maximum supply temperature of 35–45 °C).

In new buildings, for both economic and comfort reasons, a radiant heating system should be selected. These offer the additional benefit of effective use of thermal mass during possible electric cut-off times and make use of low tariff periods.

For old buildings, low-temperature radiators are generally selected. These radiators are arranged for a maximum supply temperature of 55–65 °C. A combination of both systems is also possible.

6.6 ENERGY-ECONOMIC ANALYSIS OF DIFFERENT SYSTEMS

6.6.1 Assessment of Energy Consumption for an Air-to-Water Heat Pump

Annual energy consumption of heating/cooling systems for a building contributes to minimising the life cost of the building. This consumption is obtained by time integration of instantaneous consumption during the cold season and warm season, respectively. Instantaneous consumption depends on the efficiency of the heating, ventilating and air-conditioning (HVAC) system. Computation of the annual energy consumption of a heating/cooling system can be obtained using the degree-day method or bin method [22].

6.6.1.1 Degree-Day Method

The degree-day method and its generalisations can provide a simple estimate of annual loads, which can be accurate if the indoor temperature and internal gains are relatively constant and if the heating or cooling systems operate for a complete season. The balance point temperature t_{ech} of a building is defined as that value of the outdoor temperature t_e at which, for the specified value of the indoor temperature t_i, the total heat loss is equal to the heat gain Q_{ap} from sun, occupants, lights and so forth:

$$Q_{ap} = U(t_i - t_{ech}) \tag{6.23}$$

where U is the heat transfer coefficient of the building, in W/K.

Heating is needed only when t_e drops below t_{ech}. The rate of energy consumption of the heating system is:

$$Q_{inc} = \frac{U}{\eta}[t_{ech} - t_e(\tau)]_{t_e < t_{ech}} \tag{6.24}$$

where η is the efficiency of the heating system; τ is the time.

If η, t_{ech}, and U are constant, the annual heating energy E_{inc}, in W, can be written as an integral:

$$E_{inc} = \frac{U}{\eta}\int \left[t_{ech} - t_e(\tau)\right]_+ dt \tag{6.25}$$

where the plus sign (+) below the bracket indicates that only positive values are counted.

This integral of the temperature difference conveniently summarises the effect of outdoor temperatures on a building. In practice, it is approximated by summing averages over short-time intervals (daily) and the result N_{inc}, in (K · days) is called degree-days:

$$N_{inc} = (1\text{ day})\sum_{\text{days}}(t_{ech} - t_e) \tag{6.26}$$

Here the summation is to extend over the entire year or over the heating season. The balance point temperature t_{ech} is also known as the base of the degree-days. In terms of degree-days, the annual heating energy is:

$$E_{inc} = \frac{U}{\eta}N_{inc} \tag{6.27}$$

Cooling degree-days can be calculated using an equation analogous to Eqn (6.26) for heating degree-days as:

$$N_{rac} = (1\text{ day})\sum_{\text{days}}(t_e - t_{ech}) \tag{6.28}$$

The variable-base model is used since the balance point temperature varies widely from one building to another due to widely differing personal preferences for thermostat settings and setbacks and because of different building characteristics. The basic idea is to assume a typical probability distribution of temperature data, characterised by its average $\bar{t}_{ej}$ and by its standard deviation σ. Erbs et al. [25] developed a model that needs as input only the average $\bar{t}_{ej}$ for each month of the year. The standard deviation σ_j, in °C, for each month is then estimated from the correlation:

$$\sigma_j = 1.45 - 0.029\bar{t}_{ej} + 0.0664\sigma_{an} \tag{6.29}$$

in which:

$$\sigma_{an} = \sqrt{\frac{1}{12}\sum_{j=1}^{12}(\bar{t}_{ej} - \bar{t}_{e,an})^2} \tag{6.30}$$

where σ_{an} is the standard deviation of the monthly temperature about the annual average $\bar{t}_{e,an}$.

The monthly heating degree-days $N_{\mathrm{inc},j}$ for any location are well approximated by [22]:

$$N_{\mathrm{inc},j} = \sigma_j n^{1.5}\left[\frac{\theta_j}{2} + \frac{\ln(e^{-a\theta_j} + e^{a\theta_j})}{2a}\right] \tag{6.31}$$

in which:

$$\theta_j = \frac{t_{\mathrm{ech}} - \bar{t}_{ej}}{\sigma_j\sqrt{n}} \tag{6.32}$$

where θ_j is a normalised temperature variable, n is the number of days in the month and $a = 1.698$.

The annual heating degree-days can be estimated with the equation:

$$N_{\mathrm{inc}} = \sum_{j=1}^{12} N_{\mathrm{inc},j} \tag{6.33}$$

The computer program GRAZIL has been elaborated based on a variable-base model for personal computer (PC) microsystems, using the Engineering Equation Solver (EES) software package.

6.6.1.2 Bin Method

For many applications, the degree-day method should not be used, even with the variable-base method, because the heat loss coefficient, the efficiency of the HVAC system or the balance point temperature may not be sufficiently constant. HP efficiency, for example, varies strongly with outdoor temperature t_e; the efficiency of HVAC equipment may be affected indirectly by t_e when the efficiency varies with the load (which is common for boilers and chillers). Furthermore, in most commercial buildings, occupancy has a pronounced pattern, which affects heat gain, indoor temperature and ventilation rate.

In such cases, steady-state calculations can yield good results for annual energy consumption if different temperature intervals and time periods are evaluated separately. This approach is known as the *bin method* because consumption is calculated for several values of the outdoor temperature t_e, and multiplied by the number of hours N_{bin} in the temperature interval (bin) centred on that temperature:

$$Q_{\mathrm{bin}} = N_{\mathrm{bin}}\frac{U}{1000\eta}(t_{\mathrm{ech}} - t_e)_+ \tag{6.34}$$

in which Q_{bin} is the useful energy, in kW, for each temperature interval; N_{bin} is the number of yearly hours in the temperature interval (bin) centred around the outdoor temperature; U is the heat transfer coefficient of the building, in W/K; t_{ech} is the balance point temperature, in °C; t_e is the outdoor temperature, in °C; and η is the efficiency of the HVAC system.

The subscript plus sign indicates that only positive values are considered; no heating is needed when t_e is above t_{ech} ($t_e > t_{ech}$). Eqn (6.34) is evaluated for each bin, and the total energy requirement E_{bin}, in kWh, is the sum of the Q_{bin} over all of the bins.

This method is defined in European Standard EN 15316-4.2 [26].

Knowing the capacity Q_{HP} and the drive power P_e of the HP for each bin temperature interval, the following can be determined:

- Heat demand (heat loss) of the building Q_{inc}, in kW:

$$Q_{inc} = \frac{U}{1000}(t_{ech} - t_e) \tag{6.35}$$

- HP efficiency, COP_{hp}:

$$COP_{hp} = \frac{Q_{HP}}{P_e} \tag{6.36}$$

- HP operation coefficient, f:

$$f = \min\left(1, \frac{Q_{inc}}{Q_{HP}}\right) \tag{6.37}$$

- Thermal energy provided by the HP E_t, in kWh:

$$E_t = f Q_{HP} N_{bin} \tag{6.38}$$

- Electrical energy to drive the HP E_{el}, in kWh:

$$E_{el} = f P_e N_{bin} \tag{6.39}$$

  The energy requirement E_{bin}, in kWh, is obtained by summing the values Q_{bin} given by (6.34).

- The energy delivered by the auxiliary source E_{aux}, in kWh:

$$E_{aux} = E_{bin} - E_t \tag{6.40}$$

- The total energy consumed by the HP and auxiliary source E, in kWh:

$$E = E_{el} + E_{aux} \tag{6.41}$$

The computer program METBIN has been elaborated based on this computational model in EXCEL for PC-compatible microsystems.

6.6.1.3 Numerical Application

For a building heated by an HP, the following is known: the heat transfer coefficient $U = 850$ W/K and the balance temperature $t_{ech} = 17.8$ °C. Using these, the energy consumption during the heating period is determined using the METBIN program. The results are summarised in Table 6.4.

Table 6.4 Results Provided of Computer Program METBIN

Temp. (bin) t_e (°C)	t_{ech}-t_e (°C)	Hours N_{bin} (h)	Q_{inc} (kW)	Q_{HP} (kW)	P_e (kW)	COP_{hp}	Coefficient f	E_t (kWh)	E_{el} (kWh)	E_{bin} (kWh)	E_{aux} (kWh)	E (kWh)
16	1.8	904	1.53	28.9	7.11	4.06	0.05	1383.12	340.3	1383.1	0	340.3
13	4.8	766	4.08	26.8	6.87	3.90	0.15	3125.28	801.1	3125.3	0	801.1
10	7.8	647	6.63	24.1	6.58	3.66	0.28	4289.61	1171.2	4289.6	0	1171.2
7	10.8	601	9.18	21.6	6.31	3.42	0.43	5517.18	1611.7	5517.2	0	1611.7
4	13.8	650	11.73	18.2	5.80	3.14	0.64	7624.50	2429.8	7624.5	0	2429.8
1	16.8	691	14.28	16.1	5.47	2.95	0.89	9867.48	3349.4	9867.5	0	3349.4
−2	19.8	644	16.83	14.6	5.23	2.79	1.00	9402.45	3368.1	10838.5	1430.1	4804.2
−5	22.8	497	19.38	13.3	5.01	2.65	1.00	6610.15	2490.0	9631.9	3021.8	5511.7
−8	25.8	312	21.93	12.1	4.76	2.59	1.00	3775.25	1485.1	6842.2	3067.0	4552.1
−11	28.8	162	24.48	11.6	4.66	2.49	1.00	1879.25	754.6	3965.8	2086.6	2841.5
−14	31.8	77	27.03	10.2	4.37	233	1.00	785.45	336.5	2081.3	1295.9	1632.4
−17	34.8	34	29.58	0	0	0	0	0	0	1005.7	1005.7	1005.7
−20	37.8	15	32.13	0	0	0	0	0	0	482.0	482.0	482.0
−23	40.8	5	34.68	0	0	0	0	0	0	173.4	173.4	173.4
Total								54259.47	18138.2	66827.9	12568.4	30706.0

6.6.2 Economic Analysis of Heating for a Building Using a GWHP and Other Primary Energy Sources

A study is performed here on the heating of a residential building in a rural area with a water-to-water HP, using ground-water as a heat source compared to other sources of primary energy.

6.6.2.1 Calculation Assumptions

A building with a useful area of 240 m^2 (basement-floor, ground-floor, floor and bridge) from 1993 is heated with radiators from a thermal station (TS) with gas-oil.

Indoor air temperatures were considered in accordance with the wishes of the client: 20 °C for the stairway and annex spaces; 22 °C for day rooms and bedrooms; and 24 °C for baths. The construction materials that distinguish the heated spaces are 50 cm of brick for the exterior walls, 10 cm of concrete and a 15-cm layer of expanded polystyrene insulation for the bridging, and double glazing in oak. The exterior walls will be isolated from the outside with expanded polystyrene (10 cm).

The calculation of the heat demand Q_{inc} was performed for the existing building envelope (exterior walls without insulation) and after thermal rehabilitation (exterior walls insulated with 10 cm expanded polystyrene) for different outdoor air temperatures (Table 6.5) to choose an efficient heat source.

For the DHW production, it is necessary to consider a heat demand Q_{dhw} of 3 kW (three persons, three bathrooms and a kitchen).

6.6.2.2 Description of Proposed Solution

The building heating is realised as follows:

- heating of the living spaces (living rooms, bedrooms and stairway) with floor convector-radiator
- bathroom heating with radiators (towel-port)
- hot-water temperature to radiators and convector-radiator: 50/40 °C
- the supply of radiators and convector-radiators use distributor/collector systems
- the distribution network for the radiators and convector-radiators, pexal-made, is placed at the ceiling, basement-floor, ground-floor and floor.

Table 6.5 Heat Demand for Heating

t_e (°C)	Q_{inc} (kW)	
	Actual Envelope	**Rehabilitated Envelope**
+5	18.9	13.6
0	20.2	15.5
−5	21.6	17.4
−10	23.0	18.3
−15	24.3	19.1
−20	25.6	21.1

The heat demand of the building will be provided by a Thermia Eko 180 HP and a boiler with a capacity of 300 l. A mechanical compression HP (scroll compressor) operates with the ecological refrigerant R404A. The heat source is the ground-water aquifers with a minimum temperature of 10 °C.

For the operating conditions with $t_0 = 8$ °C and $t_c = 50$ °C, the thermal power of the HP is $Q_{HP} = 21$ kW. This capacity covers part of the building's heat demand for outdoor temperatures higher than −5 °C for the existing envelope and almost the entire demand (even for an outdoor temperature of −20 °C) for the thermally-rehabilitated envelope (exterior walls with additional insulation). To meet the rest of the heat demand (i.e. heating and DHW production), the HP is equipped with three electrical resistances of 3 kW, which operate automatically, depending on the set indoor temperature. For flow rate control in the hot-water distribution network from the heating circuit, the following measures are provided:

- a first adjustment of the flow rates that supply the terminal units (radiators or convector-radiators), achieved by progressive reduction of the pipe diameters
- a base adjustment, achieved through the flow control valves for each column
- a final adjustment at the terminal units, developed by the thermostat valves set at the comfort temperature for each room.

6.6.2.3 Economic Analysis

Comparing the solution described for building heating with other possible variants of primary energy sources (liquefied petroleum gas (LPG), gas-oil and natural gas) shows a superior investment for the HP, but also shows an economy in operation costs, which enable the recovery of additional investment.

Tables 6.6 and 6.7 summarise the necessary investments and operation costs over a period of 10 years for the considered variants.

The recovery time RT of the additional investment for an HP, compared to thermal boilers using Eqn (2.43), resulted in the following:

- compared to an LPG boiler:

$$\mathrm{RT} = \frac{I_{HP} - I_{TS,LPG}}{C_{TS,LPG} - C_{HP}} = \frac{15,100 - 6500}{5033.7 - 1903.2} = 2.74 \text{ years}$$

Table 6.6 Investment Costs I, in €, for HP and Different Thermal Boilers

Solution Components	HP	Thermal Boiler with Fuel		
		LPG	Gas-Oil	Natural Gas
Heat pump/boiler	7700	3000	3000	3000
Ground-water capture	4900	–	–	–
Heat exchanger	1300	–	–	–
Circulation pumps	1200	–	–	–
Fuel tank	–	3500	3500	–
Gas connection	–	–	–	4000
Total	15,100	6500	6500	7000

Table 6.7 Operation Costs C, in €, for HP and Different Thermal Boilers

Solution Characteristics	HP	Thermal Boiler with Fuel		
		LPG	Gas-Oil	Natural Gas
Thermal power (kW)	21 + 9	24	24	24
Fuel calorific power (kW/l)	–	6.30	10.0	9.44
TS efficiency/HP coefficient of performance	2.33	0.90	0.85	0.90
Hour consumption (fuel (l/h); (m^3/h)/electricity (kW))	9.00	4.23	3.02	2.84
Annual operation (h/year)	1870[a]	1700	1700	1700
Fuel price (€/l); (€/m^3)/electricity price (€/kWh)	0.087	0.500	0.900	0.300
Annual consumption (l/year; m^3/year; kWh/year)	16,830	7191	5134	4828
Annual energy cost (€/year)	1464	3595.5	4620.5	1448.5
Estimated increase of energy price in 10 years	1.30	1.40	1.40	2.00
Operation costs (10 years), C (€)	1903.2	5033.7	6468.7	2897.0

[a]*Annual operation of electrical resistances is considered 10% of the normal operation period, so at the 1700 h/year, 170 h/year is added.*

- compared to a gas-oil boiler:

$$\mathrm{RT} = \frac{I_{\mathrm{HP}} - I_{\mathrm{TS,gas\text{-}oil}}}{C_{\mathrm{TS,gas\text{-}oil}} - C_{\mathrm{HP}}} = \frac{15,100 - 6500}{6468.7 - 1903.2} = 1.88 \text{ years}$$

- compared to a natural gas boiler:

$$\mathrm{RT} = \frac{I_{\mathrm{HP}} - I_{\mathrm{TS,natural\ gas}}}{C_{\mathrm{TS,natural\ gas}} - C_{\mathrm{HP}}} = \frac{15,100 - 7000}{2897.0 - 1903.2} = 8.15 \text{ years}$$

Compared to any of the boiler heating solutions, heating with a water-to-water HP has a recovery period of investment RT that is smaller than the normal recovery period RT_n of 8–10 years.

6.6.3 Energy, Economic and Environmental Performances of a Closed-Loop GCHP System

6.6.3.1 Description of System

The closed-loop GCHP system (Figure 6.8) represents one of the most popular configurations [27]. A working fluid is pumped through a series of vertical boreholes, where heat is collected (rejected) with a corresponding fluid temperature increase (decrease). Borehole depth is project dependent, but is usually in the 50–150-m range and the borehole-to-borehole distance ranges from 6 to 8 m.

As shown in the cross section, boreholes are usually filled with a grout to facilitate heat transfer from the fluid to the ground, and to protect ground-water aquifers. Fluid then returns to the building, where HPs either collect (reject) heat in the fluid loop, thereby decreasing (increasing) the fluid temperature. At any given time, some HPs may operate in heating mode while others might be in cooling mode. Thus, it is possible to transfer energy from one section of the building to the other via the fluid loop. Finally, in some situations it is advantageous to design the hybrid GCHP

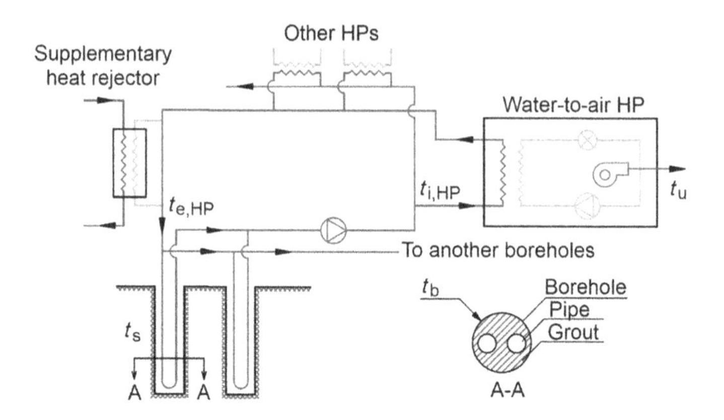

Figure 6.8 Schematic diagram of a closed-loop HGCHP system.

systems (HGCHP), where a supplementary heat rejecter or heat absorber is added to reduce the length of the GHE.

The values of COP in heating and cooling for 10 commercially available 10.5 kW water-to-air extended range HPs were considered. In cooling, the inlet fluid temperature should be as low as possible to reduce HP energy consumption. While in heating mode, the inlet fluid temperature should be as high as possible.

6.6.3.2 Analysis of System Performances

Bernier [27] considered a building which has an area of 1486 m^2 and is located in a warm-climate region, Atlanta, Georgia (a cooling-dominated climate) in the United States. This building is part of the thermal energy systems specialists (TESS) library of the TRNSYS program, and it is assumed to be equipped with 15 10.5-kW extended range HPs. The building loads were evaluated hourly using the TRNSYS simulation software and are shown in Figure 6.9. The peak building cooling load is 111 kW. The total annual building heating and cooling loads are 87,000 and 552,000 MJ, respectively. Boreholes have a 150-mm diameter and include two 25-mm HDPE-9 pipes. The borehole-to-borehole distance is set to 8 m. Bernier claimed that showing the influence of different parameters on the GHE length has determined the required length of boreholes using Eqn (5.5) for considered building and four design options:

- Case 1 uses low-efficiency HPs and a configuration borehole with a low thermal conductivity grout.
- Case 2 is similar to Case 1 except that high-efficiency HPs are used.
- In Case 3, the borehole thermal resistance has been lowered by using a high thermal-conductivity grout and spreading the pipes against the borehole wall.
- In Case 4, the GHE length has been reduced and a closed-circuit fluid cooler is used in the fluid loop (Figure 6.8).

Considering that the cooling loads are much greater than the heating loads, the GHE length was determined based on the cooling loads. It is assumed that the maximum acceptable inlet fluid temperature to the HP is 38 °C. Finally, TRNSYS simulations were used to evaluate HP energy consumption every hour over 20 years of operation, and with

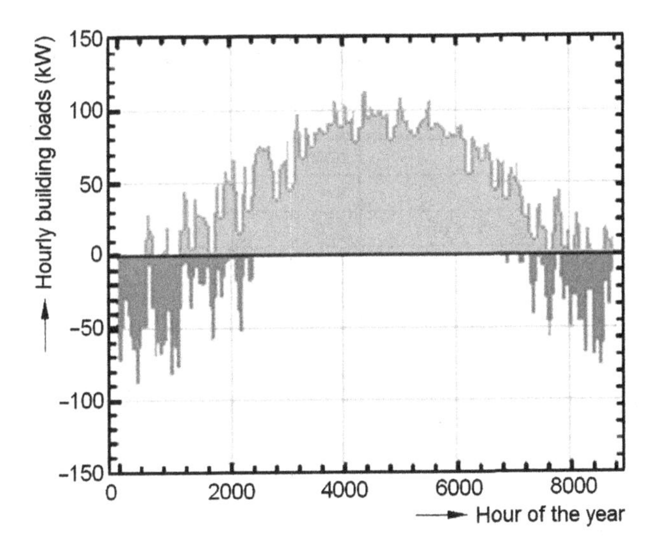

Figure 6.9 Hourly loads for considered building.

these results the average annual energy consumption and the value of the seasonal performance factor (SPF) was calculated [27]. The numerical results of length determination, summarised in Table 6.8 were examined by an analysis of energy consumption, life-cycle costs and CO_2 emissions. Main conclusions of this analysis are presented as follows:

- *Annual energy consumption.* The average annual COP for the low-efficiency HPs (Case 1) are lower than the other three cases, while the COP for Cases 2 and 3 are very similar. With the hybrid system, the ground temperatures, and consequently the inlet fluid temperature to the HPs, are higher than for Cases 2 and 3 on average. Consequently, the cooling COP for Case 4 differs from the ones observed for Cases 2 and 3 even though the same high-efficiency HPs are used.
  In terms of annual energy consumption, the low-efficiency HPs (Case 1) consume about 30% more energy than the other three cases. Cases 2 and 3 have similar energy consumption while the hybrid system consumes about 10% more energy than Cases 2 and 3. The fluid cooler of the hybrid system operates an average of 125 h per year with an average annual energy consumption of 420 kWh.
- *Life-cycle cost.* A life-cycle cost analysis is presented in Table 6.8. Numerical results show that Case 4 has the lowest life-cycle cost followed by Case 3. The main difference between these two cases has to do with borehole costs. This difference is greater than capital cost of the fluid cooler, which is estimated at 8080 € [28]. Case 2 has the lowest energy consumption followed closely by Case 3. The present value of 20 years of operation for low-efficiency HPs (Case 1) is much higher than the three other cases that use high-efficiency HPs.
- CO_2 *emissions.* The CO_2 emissions of the closed-loop GCHP system considered previously will be compared with a system that uses a gas boiler to provide heat and a conventional chiller for cooling. The hydraulic power plants not have CO_2 emissions and coal power plants present high CO_2 emissions. Figure 6.10 illustrates

Table 6.8 Comparative Numerical Results of Analysed Solutions

Specifications	Case 1	Case 2	Case 3	Case 4
Length determination of GHE				
Type of heat pump efficiency	Low	High	High	High
Hybrid system	No	No	No	Yes
Borehole thermal resistance, R_b ((m K)/W)	0.20	0.20	0.09	0.09
Bore field configuration	5×5	5×5	5×4	5×4
Total GHE length, L (m)	3165	2980	2280	1500
Annual energy consumption, in kWh				
Heating annual performance factor, SPF	4.03	5.65	5.74	5.80
Cooling annual performance factor, SPF	3.86	5.44	5.35	4.89
Heat pumps	47,730	34,440	34,760	37,580
Fluid cooler	–	–	–	420
Costs, in €				
Boreholes	79,855	75,213	63,220	41,630
Heat pumps	27,690	38,080	38,080	38,080
Fluid cooler	–	–	–	8080
Total investment cost	107,545	113,293	101,300	87,790
Operation energy cost (for 20 years)	39,160	28,252	28,514	30,830
Total costs	146,705	141,545	129,814	118,620

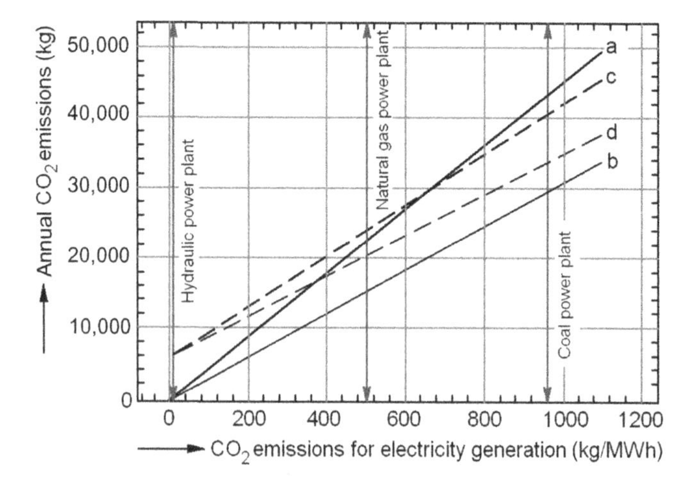

Figure 6.10 Annual CO_2 emissions as a function of CO_2 emitted per MWh of produced electricity. (a) and (b) HPs with low efficiency (Case 1) and high efficiency (Case 2), respectively; (c) and (d) gas boiler-chiller system with COPs of 4 and 5, respectively.

the amount of CO_2 emitted by these two systems for heating/cooling of the reference building. For the geothermal system, Cases 1 and 2 are considered (lines a and b). For the gas boiler-chiller system, a gas boiler efficiency of 80% is assumed and two chillers' COP (4 and 5) are considered (lines c and d).

Lines a and d intersect at 360 kg CO_2 emissions per MWh of electricity produced. Thus, if the reference building was located in a region with CO_2 emissions higher than this value, then the low-efficiency HPs will emit more CO_2 than the gas boiler and high-efficiency chiller system. A similar behaviour occurs when lines a and c intersect at 730 kg CO_2 per MWh of electricity produced. In that case the low-efficiency HPs emit more CO_2 than the gas boiler and the low-efficiency chiller system. Line b is always lower than the other three. This indicates that the operation of high-efficiency HPs leads to the least amount of annual CO_2 emissions, even in cases that utilise coal for electricity production.

6.7 PERFORMANCE ANALYSIS OF RADIATOR AND RADIANT FLOOR HEATING SYSTEMS CONNECTED TO A GCHP

6.7.1 Description of Office Room

Experimental investigations of GCHP performance were conducted in an office room (Figure 6.5) at the Polytechnic University of Timisoara, Romania, located at the ground floor of the Civil Engineering Faculty building. Timisoara has a continental temperature climate with four different seasons. The heating season runs in Timisoara from 1 October to 30 April. The following data are known: heat transfer resistance (1/*U*-value) of building components: walls (2.10 m^2 K/W), ceiling (0.34 m^2 K/W), windows and doors (0.65 m^2 K/W); glass walls surface, 8.2 m^2; total internal heat gain (e.g. from computers, human and lights), 25 W/m^2; and heat demand, 1.35 kW. The indoor and outdoor air design temperatures are 22 and −15 °C, respectively.

This space is equipped both with a floor heating system and steel panel radiators to analyse the energy and environmental performances of these systems. These two heating systems are connected to a mechanical compression GCHP, type WPC 5 COOL. In the GCHP system, heat is extracted from the ground by a closed-loop vertical GHE with a length of 80 m. Figure 6.11 illustrates the monthly energy demand for office room heating.

6.7.2 Experimental Facilities

The GCHP experimental system consisted of a BHE, HP unit, circulating water pumps, floor/radiator heating circuit, data acquisition instruments and auxiliary parts as shown in Figure 6.12.

6.7.2.1 Borehole Heat Exchanger

The GHE of this experimental GCHP consisted of a simple vertical borehole that had a depth of 80 m. Antifreeze fluid (30% ethylene glycol aqueous solution) circulates in a single polyethylene U-tube of 32 mm internal diameter, with a 60-mm separation between the return and supply tubes, buried in the borehole. The borehole's overall diameter was 110 mm. The borehole was filled with sand and finished with a bentonite layer at the top to avoid intrusion of pollutants in the aquifers. The average temperature across the full borehole depth tested was 15.1 °C.

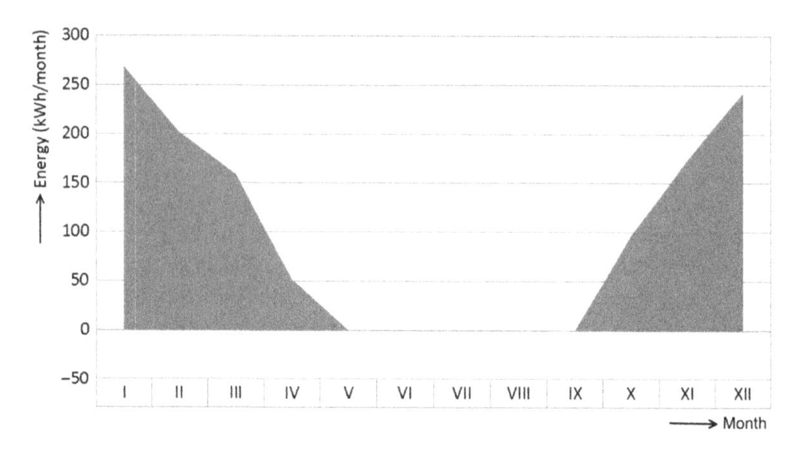

Figure 6.11 Monthly energy demand for office room heating.

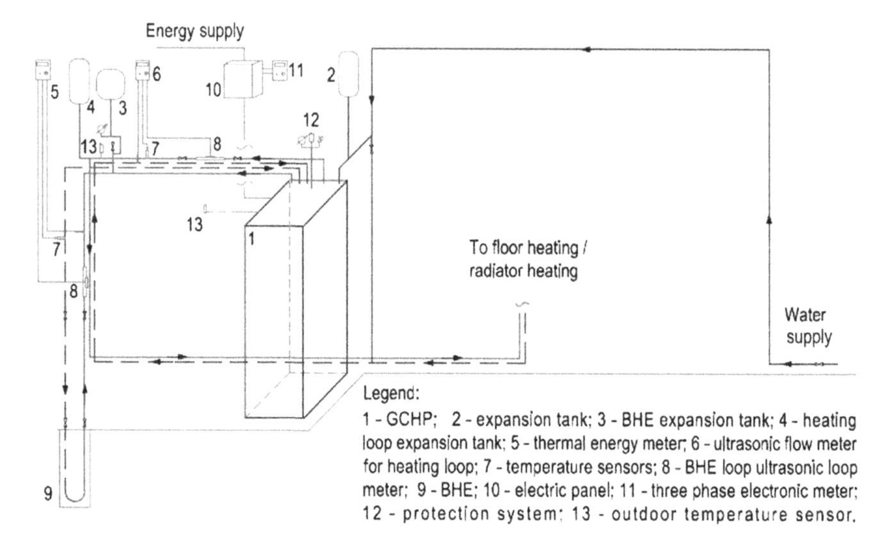

Figure 6.12 Experimental GCHP system.

The ground characteristics are based on measurements obtained from the Banat Water Resources Management Agency [29]. The average thermal conductivity and thermal diffusivity of the ground from the surface to 80 m deep tested were 1.90 W/(m K) and 0.79×10^{-6} m^2/s, respectively [30]. The boreholes were completely backfilled with grout mixed with drilling mud, cement and sand in specific proportions. The thermal conductivity and thermal diffusivity of the grout tested by the manufacturer were 2.32 W/(m K) and 0.93×10^{-6} m^2/s, respectively.

6.7.2.2 Heat Pump Unit

The HP unit is a reversible ground-to-water scroll hermetic compressor unit with R410A as a refrigerant and the nominal heating capacity of 6.5 kW. The HP unit is a compact type model that has an inside refrigeration system. The operation of the HP is governed by an electronic controller, which, depending on the system water return temperature, switches the HP compressor on or off. The heat source circulation pump was controlled by the HP controller, which activates the source pump 30 s before compressor activation.

6.7.2.3 GCHP Data Acquisition System

The GCHP data acquisition system consists of the indoor and outdoor air temperature, dew point temperature, supply/return temperature, heat source temperature (outlet BHE temperature), relative air humidity and the main operating parameters of the system components.

6.7.2.4 Heating Systems

The heating systems are supplied via a five-circuit flow/return manifold as follows. The first two circuits supply the floor heating system. The third and fourth circuits are coupled to a radiator heating system, and the fifth circuit is for backup.

The flow/return manifold is equipped with a circulation pump to ensure the chosen temperature of the heat carrier (hot water). A three-way valve and a thermostatic valve are provided to adjust the maximum hot-water temperature of the floor's heating system. Thus, for higher temperatures, the hot water is adjusted to achieve a circulation loop in the heating system.

To achieve higher performances of the heating systems, a thermostat is provided for controlling the start/stop command of the circulation pump when the room reaches the set point temperature. At the same height as this thermostat, there is also an ambient thermostat that controls the starting and stopping of the HP to ensure optimum operation of the entire heating system.

The start-stop command of the flow/return manifold circulation pump is controlled by an interior thermostat relay, situated at a height of approximately 1.00 m above the floor surface. This height has been determined to provide adequate comfort for the office occupants.

The *radiant floor heating system* consists of two circuits connected to a flow/return manifold (Figure 6.13), designed to satisfy the office heating demand of 1.35 kW. The first circuit has a length of 54 m and is installed in a spiral coil, with the closest step distance to the exterior wall of the building to compensate for the effect of the heat bridge, and the second circuit, with a length of 61 m, is mounted in the coil simple. The mounting step of the coils is between 10 and 30 cm.

The floor heating pipes are made of cross-linked polyethylene with an external diameter of 17 mm and a wall thickness of 2 mm. The mass flow rate for each circuit is controlled by the flow/return manifold circuit valves. They are adjusted to satisfy the heat demand according to Timisoara's climate ($t_e = -15$ °C).

Radiator heating system. The low-temperature radiator heating system (45/35 °C) has two steel panel radiators, each one with two water columns and a length of 1000 mm, height of 600 mm and thermal power of 680 W (Figure 6.14), connected to a flow/return manifold and dimensioned to satisfy the office heating demand of 1.35 kW. They are installed on a stand at 15 cm above the floor surface to ensure optimal indoor air circulation.

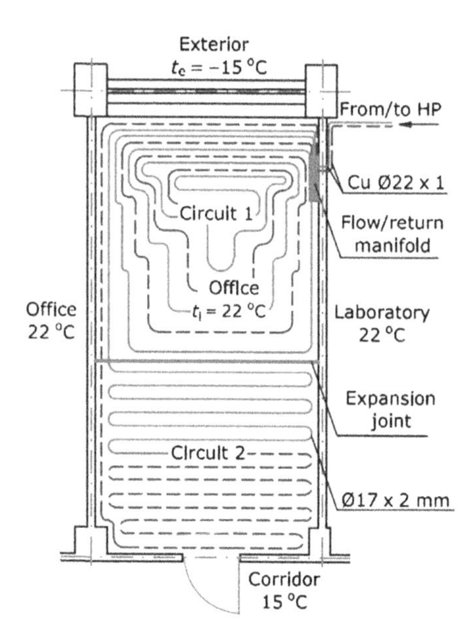

Figure 6.13 Schematics of floor heating circuit.

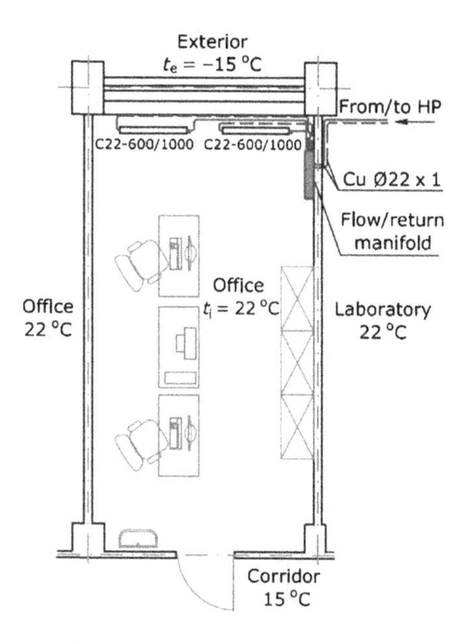

Figure 6.14 Schematics of radiator heating system.

The heating radiator system pipes are made of cross-linked polyethylene with an external diameter of 17 mm and a wall thickness of 2 mm. The mass flow rate for each radiator is controlled by the flow/return manifold circuit valves, adjusted to satisfy the heat demand of the office room.

6.7.3 Measuring Apparatus

A network of sensors was set up to allow monitoring of the most relevant parameters of the system [31]. Two thermal energy meters were used to measure the thermal energy produced by the GCHP and the extracted/injected thermal energy to the ground. A thermal energy meter was built with a heat computer, two PT500 temperature sensors and an ultrasonic mass flow meter. The two PT500 wires temperature sensors with an accuracy of $\pm 0.15\,^{\circ}\mathrm{C}$ were used to measure the supply and return temperature for a hydraulic circuit (the water-antifreeze solution circuit or the manifold circuit). Also, an ultrasonic mass flow meter measured the mass flow rate for a hydraulic circuit. The thermal energy meters were AEM meters, model LUXTERM, with a signal converter IP 67 and accuracy <0.2%. A three-phase electronic electricity meter measured the electrical energy consumed by system (the HP unit, the circulating pumps, a feeder 220 Vca/24 Vcc, a frequency converter and a programmable logic controller) and another three-phase electronic electricity meter measured the electrical energy consumed by the HP compressor. The two three-phase electronic electricity meters were multifunctional type from AEM, model ENERLUX-T, with an accuracy grade in $\pm 0.4\%$ of the nominal value. The monitoring and recording of the experiments were performed using a PC. The indoor and outdoor air temperature was measured by air flow sensors and supply/return and heat source temperature were recorded by positive temperature coefficient (PTC) immersion sensors, all connected to the GCHP data acquisition system and having an accuracy of $\pm 0.2\,^{\circ}\mathrm{C}$.

6.7.4 Experimental Results

6.7.4.1 Comparison Between Energy Performances of Systems

The two heating systems were monitored for 2 months. The experiments were conducted for a 1-week heating period for each of the two analysed heating systems, from the 7th of December 2013 to the 6th of January 2014 and from the 15th of January 2014 to the 14th of February 2014. The outdoor temperature varied in the range of -5.6 to $9.7\,^{\circ}\mathrm{C}$. The weekly mean values of the outdoor temperature during the two periods were almost equal.

The energy performance of heating system is determined based on the coefficient of performance (COP_{sys}), which can be calculated using Eqn (2.12).

The carbon dioxide emission (C_{CO_2}) of the heating system during its operation is calculated with Eqn (2.49).

To obtain the COP and CO_2 emissions, it is necessary to measure the heating energy and electricity used in the system.

During the cold season, measurements were performed at the appreciatively same average outdoor air temperature and the heat source temperature for both the radiant floor heating system and the radiator heating system. The following average values were recorded: outdoor air temperature (t_e), indoor air temperature (t_i), heat source

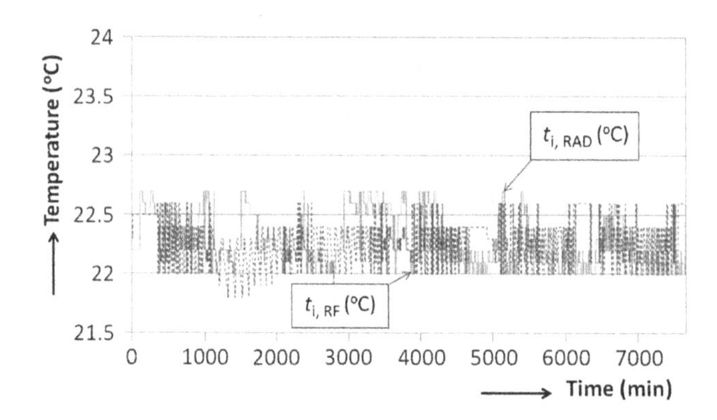

Figure 6.15 Variation of indoor air temperature.

Table 6.9 Experimental Results

Heating System	t_e (°C)	t_i (°C)	t_{hs} (°C)	t_d (°C)	E_{el} (kWh)	E_t (kWh)	C_{CO_2} (kg)	ON/OFF Switching	COP_{sys}
Radiant floor	9.39	22.28	18.77	28.12	5.77	32.78	3.16	48	5.68
Radiator	9.00	22.30	17.62	30.62	6.35	34.42	3.47	140	5.42

temperature (t_{hs}), supply hot-water temperature (t_d), electricity consumption (E_{el}) and useful thermal energy for heating (E_t). In addition, the CO_2 emission and the ON/OFF switching of the HP were determined in both heating systems.

Figure 6.15 shows a comparison between the indoor air temperatures $t_{i,RAD}$ and $t_{i,RF}$ obtained by radiator heating and radiant floor heating. It is observed that due to the small thermal inertia of the radiators, a high level of ON/OFF switching is needed for the HP of the radiator heating system, leading to large fluctuations of indoor air temperature compared with the floor heating system, along with reduced thermal comfort. Table 6.9 presents a summary of the experimental results.

The two heating systems have small differences (4.5%) in their energy performance coefficient (COP_{sys}) value, but the ON/OFF switching in the case of radiator heating system is almost three times higher than that for radiant floor heating system, leading to higher wear on the HP equipment. In addition, there was 10% higher energy consumption and CO_2 emission for the radiator heating system, compared with the floor heating system under the same operating conditions.

Energy consumption can be influenced by building occupants' activity and the floor surface material. If the floor surface material exhibits good heat transfer, such as with stone or tile, the floor feels cold even at a temperature of approximately 24–25 °C.

Generally, the building occupants want the floor to feel warm to the feet, and this is why they increase the water temperature to a level that makes the floor feel warm, sometimes even in summer. The warm temperature is typically more than 27 °C for stone-based materials. The excess heat must be ventilated/cooled to retain acceptable indoor air temperatures. This causes a great increase in energy consumption.

Cases in which the energy consumption has doubled have been observed in studies. In a well-insulated building, the selected floor surface material is of crucial importance in regard to how warm the floor feels. For example, oak parquet at a temperature of 21 °C and stone floor at a temperature of 26 °C feel neutral and roughly the same under a bare foot according ISO/TS 13732-2 [32]. However, this is not always the case, the percent dissatisfied (PD) in % has a relation with floor surface temperature as follow [33]:

$$PD = 100 - 94\exp(-1.387 + 0.118t_{f,m} - 0.0025t_{f,m}^2) \tag{6.42}$$

where t_f is the mean floor surface temperature.

6.7.4.2 Uncertainty Analysis

Uncertainty analysis (the analysis of uncertainties in experimental measurement and results) is necessary to evaluate the experimental data. An uncertainty analysis was performed using the method described by Holman [34]. A result Z is a given function of the independent variables $x_1, x_2, x_3 \ldots x_n$. If the uncertainties in the independent variables $w_1, w_2, w_3 \ldots w_n$ are all given with same odds, then uncertainty in the result w_Z having these odds is calculated by the following equation [34]:

$$w_Z = \sqrt{\left(\frac{\partial Z}{\partial x_1} w_1\right)^2 + \left(\frac{\partial Z}{\partial x_2} w_2\right)^2 + \cdots + \left(\frac{\partial Z}{\partial x_n} w_n\right)^2} \tag{6.43}$$

In the present study, the temperatures, thermal energy and electrical energy were measured with appropriate instruments explained previously. Error analysis for estimating the maximum uncertainty in the experimental results was performed using Eqn (6.43). It was found that the maximum uncertainty in the results is in the COP_{sys}, with an acceptable uncertainty of 3.9 and 3.1% for radiant floor heating system and radiator heating system, respectively.

6.7.5 Thermal Comfort Assessment

The office room with geometrical dimensions from Figure 6.5 is considered. The following data are known: indoor air temperature, 22 °C; relative humidity of air, 55%; thermal power of heater, 1360 W; floor temperature, 20 °C for radiator heating and 29 °C for radiant floor heating.

Assessment of thermal comfort in the office room is performed using the predicted mean vote (PMV)–predicted percent dissatisfied (PPD) model [13]. A comparative study of PMV and PPD indices is performed using the computer program THERMAL COMFORT [9] in several points situated on a straight line (discontinuous), at different distances from the window, function of metabolic rate (i_M) and clothing thermal resistance (R_{cl}).

The results of the numerical solution obtained for the pairs of values are as follows: 3.4 met-0.67 clo (intense activity, normal clothes), 1 met-0.90 clo (reading seated, winter clothes) and 1.1 met-0.29 clo (writing, light clothes), and are reported in Table 6.10.

According to the performed study, it was established that the PMV index has values close to zero only for the pair of values 1 met-0.9 clo. For any other pair of values $i_M - R_{cl}$, the percent of people dissatisfied with their thermal comfort would be greater than 35%. In addition, the PMV index values for the pair 1 met-0.9 clo are lower with 47–94% in the case of the radiant floor heating system than in the case of the radiator heating system. Therefore, the first system leads to increased thermal comfort.

Table 6.10 Numerical Results of THERMAL COMFORT Computer Program

Heating Type	Distance from the Window (m)	3.4 met-0.67 clo			1 met-0.90 clo			1.1 met-0.29 clo		
		t_r (°C)	PMV (–)	PPD (%)	t_r (°C)	PMV (–)	PPD (%)	t_r (°C)	PMV (–)	PPD (%)
Radiant floor	1.0	23.00	2.17	84	23.00	− 0.35	8	23.00	− 1.63	58
	1.5	23.70	2.22	86	23.70	− 0.26	6	23.70	− 1.51	52
	2.0	24.30	2.26	87	24.30	− 0.18	6	24.30	− 1.41	46
	2.5	24.70	2.28	88	24.70	− 0.12	5	24.70	− 1.34	42
	3.0	25.00	2.31	88	25.00	− 0.08	5	25.00	− 1.28	39
	3.5	25.20	2.32	89	25.20	− 0.06	5	25.20	− 1.25	38
	4.0	25.30	2.32	89	25.30	− 0.04	5	25.30	− 1.23	37
	4.5	25.50	2.34	89	25.50	− 0.02	5	25.50	− 1.19	35
	5.0	25.50	2.34	89	25.50	− 0.02	5	25.50	− 1.19	35
Radiator	1.0	20.60	2.01	77	20.60	− 0.67	14	20.60	− 2.05	79
	1.5	21.20	2.05	79	21.20	− 0.59	12	21.20	− 1.94	74
	2.0	21.70	2.08	80	21.70	− 0.53	11	21.70	− 1.86	70
	2.5	22.10	2.11	82	22.10	− 0.48	10	22.10	− 1.79	67
	3.0	22.40	2.13	82	22.40	− 0.43	9	22.40	− 1.74	64
	3.5	22.60	2.14	83	22.60	− 0.41	8	22.60	− 1.70	62
	4.0	22.70	2.15	83	22.70	− 0.39	8	22.70	− 1.69	61
	4.5	22.80	2.16	83	22.80	− 0.38	8	22.80	− 1.67	60
	5.0	22.80	2.16	83	22.80	− 0.38	8	22.80	− 1.67	60

6.7.6 Numerical Simulation of Useful Thermal Energy and System COP Using TRNSYS Software

TRNSYS software [35] is one of the most flexible modelling and simulation tools and can solve very complex problems from the decomposition of the model in various interconnected model components. One of the main advantages of TRNSYS for the modelling and design of GSHPs is that it includes components for the calculation of building thermal loads, specific components for HVAC, HPs and circulating pumps, modules for BHEs and thermal storage, as well as climatic data files, which make it a very suitable tool to model a complete A/C HP installation to provide heating and cooling to a building.

The simulation error can be estimated by the RMS, c_v, R^2 values computed with Eqns (6.20)–(6.22), respectively and the percentage difference (relative error) e_r which is defined as:

$$e_r = \frac{|y_{mea,i} - y_{sim,i}|}{y_{mea,i}} 100\% \tag{6.44}$$

where $y_{mea,i}$ is the measured value; $y_{sim,i}$ is the potential value at point i obtained by simulation.

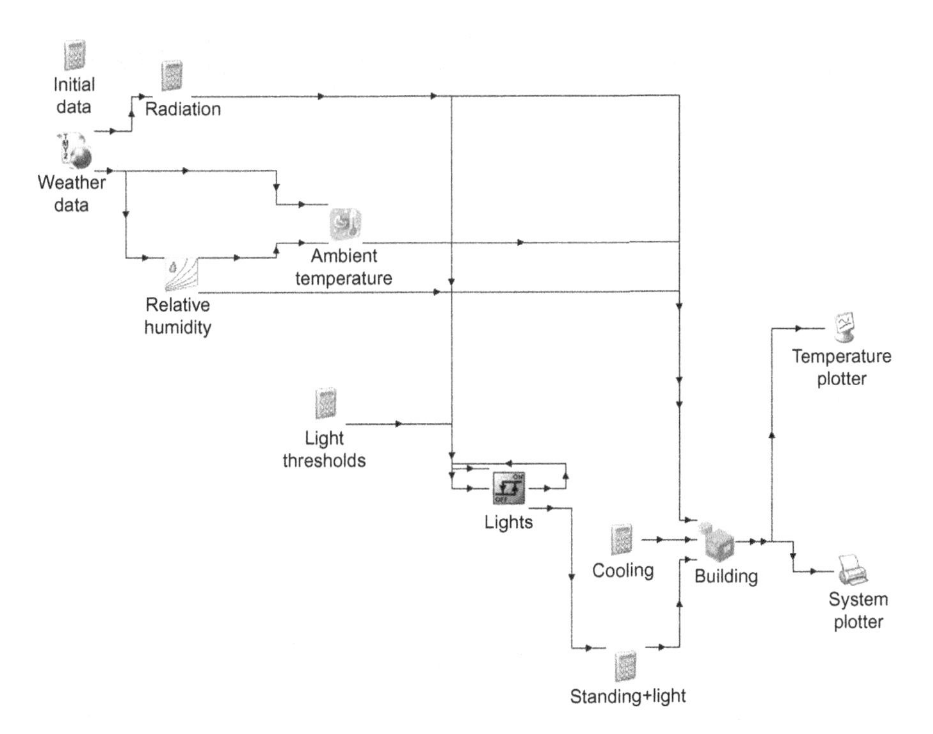

Figure 6.16 Scheme of the system model built in TRNSYS to simulate useful thermal energy.

6.7.6.1 Simulation of Thermal Energy Used for Office Room Heating

Definition of the operation scheme. To simulate the thermal energy used to cover the heating load of the office room, the operational connections were established between the building and all internal and external factors.

Figure 6.16 presents the operational scheme built in TRNSYS, where the building thermal behaviour was modelled using a 'Type 56' subroutine. This subroutine was processed with the TRNBuild interface by introducing the main construction elements, their orientation and surface, shadow factors, and indoor activity type. Weather data for Timisoara were obtained from the Meteonorm database [36] and the weather data reader 'Type 109' and 'Type 89d' were used to convert the data to a form readable from TRNSYS.

The simulation model took into account the outdoor air infiltrations, heat source type, and interior gains. To extract the results, an online plotter ('Type 65') is used.

Simulation results and comparison with experimental data. Performing simulations for a 1-year period (8760 h), the values of thermal energy used for heating were obtained and are presented beside the measured values in Table 6.11. Statistical values such as RMS, c_v and R^2 are also given in Table 6.11.

There was a maximum difference between the measured and TRNSYS simulated values for the heating period of approximately 1.59%, which is very acceptable. The RMS and c_v values in heating mode are 2.722 and 1.41%, respectively. The R^2-values are about 0.9999, which can be considered as very satisfactory. Thus, the simulation model was validated by the experimental data.

Table 6.11 Thermal Energy Used for Office Room Heating

Month	Heating Energy (kWh)		Percentage Difference e_r (%)	RMS	c_v (%)	R^2
	Simulated	Measured				
January	252.50	256.24	1.57	2.72187	1.409	0.99990075
February	195.70	195.06	0.32			
March	151.61	150.44	0.77			
April	49.73	48.95	1.59			
May	0.00	0.00	0.00			
June	0.00	0.00	0.00			
July	0.00	0.00	0.00			
August	0.00	0.00	0.00			
September	0.00	0.00	0.00			
October	94.85	95.66	0.84			
November	174.45	172.62	1.06			
December	238.75	240.11	0.57			

6.7.6.2 COP Simulation of GCHP System

Definition of the operation scheme. For COP simulation of the GCHP system, the operational scheme built in TRNSYS from Figure 6.17 was utilised. The assembly of GCHP system consists of the standard TRNSYS weather data readers 'Type 15-6', a GCHP model 'Type 919', a BHE 'Type 557a'. Also, in the simulation model were defined single-speed circulating pumps 'Type 114' for the antifreeze fluid in the BHE and 'Type 3d' for heat carrier fluid of the manifold. A 'Type 14' for the load profile and a daily load subroutine were created, this approach improving significantly the numerical convergence of the model. Finally, two model integrators ('Type 25' and 'Type 24') were used to calculate daily and total results for thermal energy produced.

Simulation results and comparison with experimental data. COP simulation of the GCHP integrated both with radiator and radiant floor heating systems was performed for a 1-month period. The results of the simulation program are presented beside the experimental data in Table 6.12. A comparative analysis of these results indicates that the COP_{sys} values simulated with TRNSYS program were only 3.52% lower than the measured values for radiant floor heating system and only 4.98% lower than the measured values for radiator heating system. Thus, the simulation model is validated experimentally.

6.7.7 Conclusions

The use of HPs in modern buildings with improved thermal insulation and reduced thermal load is a good alternative to traditional heating solutions.

This study showed that radiator heating and radiant floor heating systems have small differences (4.5%) in their energy performance coefficient (COP_{sys}) value, but the ON/OFF switching in the case of a radiator heating system is almost three times higher than that for a radiant floor heating system, leading to higher wear on the HP

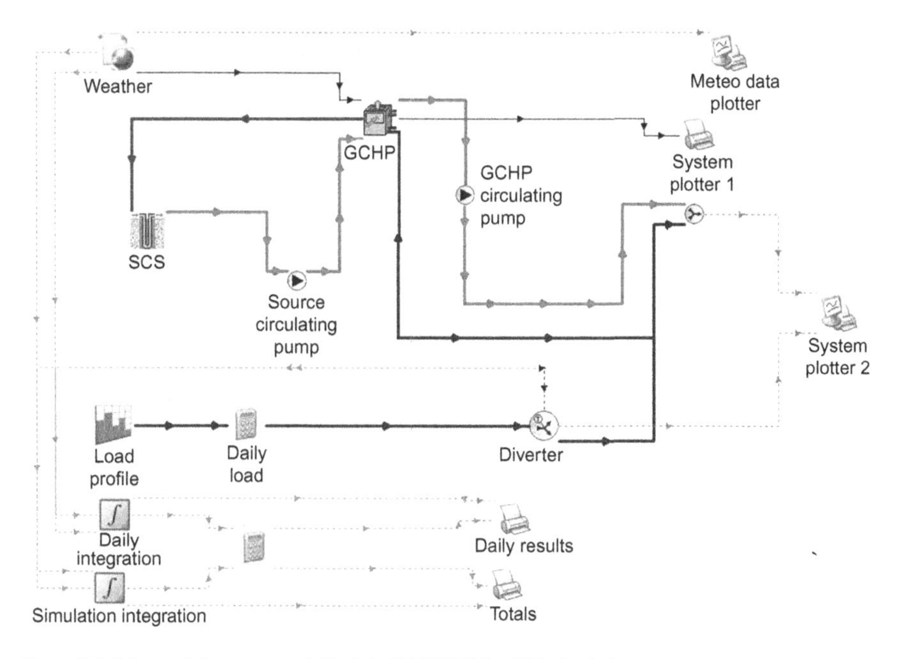

Figure 6.17 Scheme of the system model built in TRANSYS for COP simulation.

Table 6.12 The COP Values for GCHP System

Heating System	COP_{sys}		Percentage Difference
	Simulated	Measured	e_r (%)
Radiant floor	5.48	5.68	3.52
Radiator	5.15	5.42	4.98

equipment. In addition, the radiator heating system showed 10% higher energy consumption and CO_2 emissions compared to the floor heating system under the same operating conditions.

The developed TRNSYS simulation models can be used as a tool to determine the GCHP performance connected with different heating systems to optimise their energy efficiency and ensure the user's comfort throughout the year.

6.8 CONTROL OF HEAT PUMP HEATING AND COOLING SYSTEMS

6.8.1 Control System

The control of a heating and cooling system can be classified as central control, zone control and individual room control [17]. Figure 6.18 is a diagram on the principles of control [20].

The central control controls the supply water temperature for the heating system based on the outdoor air temperature. The room control then controls the water flow rate or water temperature for each room according to the room set-point temperature.

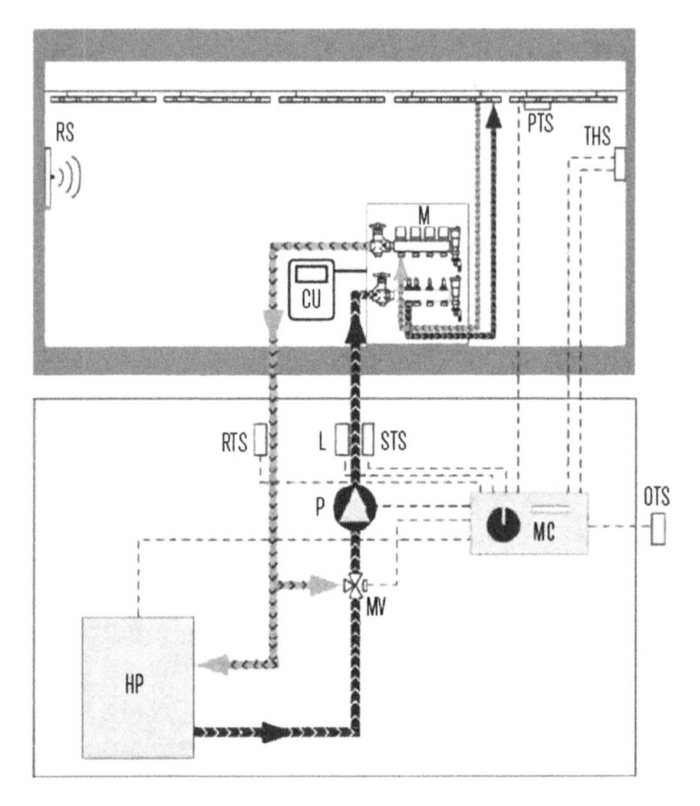

Figure 6.18 Diagram of the principles of control exemplified by a radiant floor system. HP, heat pump; CU, control unit; PTS, panel temperature sensor; L, limiter; M, manifold; MC, main controller; MV, mixing valve; OTS, outdoor temperature sensor; P, pump; RS, room sensor; RTS, return medium temperature sensor; STS, supply medium temperature sensor; THS, temperature-humidity sensor.

Instead of controlling the supply water temperature, it is recommended to control the average water temperature (mean value of supply and return water temperature) according to outdoor and/or indoor temperatures. During the heating period, as the internal load increases, the heat output from the heating system will decrease and the return temperature will rise. If the control system controls the average water temperature, the supply water temperature will automatically decrease due to the increased return water temperature. This results in a faster and more accurate control of the thermal output to the space and will give better energy performance than controlling the supply water temperature.

Radiant surface cooling systems need controls to avoid condensation. This can be done by a central control of the supply water temperature limiting the minimum water temperature based on the zone with the highest dew-point temperature. If the supply water temperature is limited, the temperature of the rest of system will be higher than the dew-point, and there is no risk of condensation on the pipes, and on the surface of the radiant system. Limiting the supply water temperature will lower the cooling power of a radiant system at high indoor humidity levels.

Larger buildings should be divided in several different thermal zones to optimise energy and control performance. Each zone can be controlled with reference to a temperature sensor in a representative space of the zone.

For the improved comfort and further energy savings, use an individual room temperature control. Each valve on the manifold is controlled by each room thermostat. An apartment or one-family house was normally regarded as one zone, but installing thermostats for each room is becoming popular. For better thermal comfort, it is preferable to control the room temperature as a function of the operative temperature.

The heat capacity of surfaces with embedded pipes plays a significant role for the thermodynamic properties of the heating system and, hence, for the control strategy. An obvious consequence of the response time of a conventional floor structures is that the instant control of the heating power is not necessary. The temperature of heat carrier, the time response and the thermal capacity of systems depends on the thickness of the surface layer where the pipes are embedded.

For a low-temperature heating and high-temperature cooling system, a significant effect is the 'self-regulating' control [17]. This self-regulating effect depends partially on the temperature difference between the room and a heated/cooled surface, and partly on the difference between the room and the average water temperature in the embedded pipes. This impact is greater for systems with surface temperatures close to a room temperature because the small temperature change represents a higher percentage compared to the same temperature change at a high temperature difference. The self-regulating effect supports the control equipment in maintaining a stable thermal environment, and providing comfort to the occupants in the room.

6.8.2 Heat Pump Control

A properly planned heat distribution system is designed to release heat in a manner that reduces large temperature swings in living spaces. In large HP systems, stepped control is common and the auxiliary systems (fans, circulation pumps, etc.) should operate in a similar manner to maintain the highest possible coefficients of performance. A further possibility is the use of a frequency converter to allow for variable drive control. In this case, proper lubrication of the compressor must be ensured.

Depending on the manufacturer's configuration, aside from the control of supply and return temperatures, the controller must achieve the following tasks [37]:

- limit start/stop frequency (start delay)
- defrost the evaporator for air-to-water HPs
- optional DHW heating
- optional mixing control
- additional heat source control; and
- safety functions.

For bivalent and interruptible monovalent operation HPs, temporarily interrupting the power supply must be possible. For this reason, two separate electrical circuits are required.

REFERENCES

[1] Allard F, Seppänen O. European actions to improve energy efficiency of buildings. Rehva J 2008;45(1):10–20.
[2] Ellison RD. The effects of reduced indoor temperature and night setback on energy consumption of residential heat pumps. ASHRAE Trans 1977;84(2):352–63.
[3] Sarbu I, Kalmar F, Cinca M. Thermal building equipments – energy optimization and modernization. Timisoara: Polytech Publishing House; 2007 [in Romanian]
[4] Sarbu I, Dan D, Sebarchievici C. Performances of heat pump systems as users of renewable energy for building heating/cooling. WSEAS Trans Heat Mass Transf 2014;9:51–62.
[5] Lund JW. Direct-use of geothermal energy in the USA. Appl Energy 2003;74:33–42.
[6] Yang W, Zhou J, Xu W, Zhang GQ. Current status of ground-source heat pumps in China. Energy Policy 2009;38(1):323–32.
[7] Lee JY. Current status of ground source heat pumps in Korea. Renewable Sustainable Energy Rev 2009;13:1560–8.
[8] Sarbu I, Sebarchievici C. General review of ground-source heat pump system for heating and cooling of buildings. Energy Build 2014;70(2):441–54.
[9] THERMAL COMFORT tool, Version 2. Berkeley, CA: ASHRAE, Centre for the Built Environment; 2011.
[10] BUDERUS. Handbuch fur Heizung-stechnik. Berlin: Beuth Verlag; 1994.
[11] Ilina M, Burchiu S. Influence of heating systems on microclimate from living rooms. Fitter, Romania 1996;6:24–9.
[12] Berglund LG, Fobelets A. A subjective human response to low level air currents and asymmetric radiation. ASHRAE Trans 1987;93(1):497–523.
[13] Sarbu I, Sebarchievici C. Aspects of indoor environmental quality assessment in buildings. Energy Build 2013;60(5):410–19.
[14] Hesaraki A, Holmberg S. Energy performance of low temperature heating systems in five new-built Swedish dwellings: a case study using simulations and on-site measurements. Build Environ 2013;64:85–93.
[15] ASHRAE handbook, HVAC systems and equipment. Atlanta, GA: American Society of Heating, Refrigerating and Air Conditioning Engineers; 2012.
[16] REHVA, Guidebook no 7: Low temperature heating and high temperature cooling; 2007.
[17] Kim KW, Olesen BW. Radiant heating and cooling systems. ASHRAE J 2015;57(2,3):28–37 34–42
[18] ISO 7730. Moderate thermal environment – determination of the PMV and PPD indices and specification of the conditions for thermal comfort. Geneva: International Organization for Standardization; 2005.
[19] Sarbu I, Sebarchievici C. A study of the performances of low-temperature heating systems. Energ Effic 2015;8(3):609–27.
[20] ISO 11855. Building environment design – Design, dimensioning, installation and control of the embedded radiant heating and cooling systems. Geneva: International Organization for Standardization; 2012.
[21] Roumajon J. Modélisation numerique des émissions thermiques. Chaud, Froid Plomberie 1996;579(4):55–8.
[22] ASHRAE handbook. Fundamentals. Atlanta, GA: American Society of Heating, Refrigerating and Air-Conditioning; 2013.
[23] ASHRAE handbook, HVAC applications. Atlanta, GA: American Society of Heating, Refrigerating and Air–Conditioning Engineers; 2011.
[24] Bechthler H, Browne MW, Bansal PK, Kecman V. New approach to dynamic modelling of vapour-compression liquid chillers: artificial neural networks. Appl Therm Eng 2001;21 (9):941–53.
[25] Erbs DG, Klein SA, Beckman WA. Estimation of degree-days and ambient temperature bin data from monthly-average temperatures. ASHRAE J 1983;25(6):60.
[26] Zirngib J. Standardization activities for heat pumps. Rehva J 2009;46(3):24–9.
[27] Bernier M. Closed-loop ground-coupled heat pump systems. ASHRAE J 2006;48(9):13–24.
[28] Yavuzturk C, Spitler JD. Comparative study of operating and control strategies for hybrid ground-source heat pump systems using a short time step simulation model. ASHRAE Trans 2000;106(2):192–209.

[29] Banat Water Resources Management Agency, Records for soil properties data of Timisoara, Romania; 2011.
[30] Sebarchievici C. Optimization of thermal systems from buildings to reduce energy consumption and CO_2 emissions using ground-coupled heat pump [Doctoral thesis]. Romania: Polytechnic University Timisoara; 2013.
[31] Sebarchievici C, Sarbu I. Performance of an experimental ground-coupled heat pump system for heating, cooling and domestic hot-water operation. Renewable Energy 2015;76(4):148–59.
[32] ISO/TS 13732-2. Ergonomics of the thermal environment. Methods for the assessment of human responses to contact with surface, Part 2: Human contact with surfaces at moderate temperature. Geneva: International Organization for Standardization; 2001.
[33] ASHRAE Standard 55. Thermal environmental conditions for human occupancy. Atlanta, GA: American Society of Heating, Refrigerating and Air-Conditioning Engineers; 2010.
[34] Holman JP. Experimental method for engineers. Singapore: McGraw Hill; 2001.
[35] TRNSYS 17. A transient system simulation program user manual. USA: Solar Energy Laboratory, University of Wisconsin-Madison; 2012.
[36] METEONORM. Help, Version 5.1; 2004.
[37] Ochsner K. Geothermal heat pumps: a guide to planning & installing. London-Sterling. Earthscan; 2007.

CHAPTER 7

Experimental Ground-Coupled Heat Pump Systems

7.1 GENERALITIES

A number of ground-coupled heat pump (GCHP) systems have been used in residential and commercial buildings worldwide because of their high efficiency and environmental friendliness [1–4]. Due to the heat capacity of the ground, ambient air temperature variations are directly reflected only in the surface ground temperature, and their effect is reduced at deeper layers. According to the reports of the 2010 World Geothermal Congress, GCHP systems have the largest energy use and installed capacity, accounting for 69.7% and 49.0% of the worldwide capacity and use, respectively. The installed capacity of GCHPs is 35,236 MWt, and their annual energy use is 214,782 TJ/year, with a capacity factor of 0.19 (in the heating mode). Almost all of the installations occur in North America, Europe and China, increasing from 26 countries in 2000, to 33 countries in 2005 and all the way up to the present 43 countries. Sweden, Denmark, Switzerland, Austria and the United States are the leaders in this field [5]. The number of installed GCHP systems has grown continuously by 10–30% annually in recent decades [2,6,7]. Extrapolating currently observed growth rates for Europe of 5.4 million heat pump (HP) units per year leads to an expectation of 70 million installed units in Europe by 2020 [3]. The use of GCHPs to achieve adequate temperatures has been studied by several researchers [8,9].

A GCHP system consists of an HP unit coupled with a ground heat exchanger (GHE), usually a vertical borehole heat exchanger (BHE) or, less commonly, horizontal loops. A closed single or double U-tube is often inserted inside the borehole and a heat carrier fluid is circulated in the U-tube to exchange heat with the surroundings. For safety and stability reasons, a bentonite-cement suspension or an enhanced-cement is used to backfill the space between the U-tube and its surrounding soil or rock. A GCHP uses the ground as a heat source in heating and a heat sink in cooling. In the heating mode, a GCHP absorbs heat from the ground and uses it to heat the building. In the cooling mode, heat is absorbed from the conditioned spaces and transferred to the earth through the GHE.

Most existing studies of GCHP systems concentrate on theoretical and simulation model research [2,10,11] or *in situ* monitoring of heat transfer in the BHE [6,7,12–14]. Only a few researchers have investigated the experimental operational performance of GCHP systems. Hwang et al. [15] presented the actual cooling performance of a GCHP system installed in Korea for 1 day of operation. Pulat et al. [16] evaluated the performance of a GCHP with a horizontal GHE installed in Turkey under winter climatic conditions. The COP of the entire system and the HP unit were found to be 2.46–2.58 and 4.03–4.18, respectively. Yang et al. [17] reported the heat transfer of a

Ground-Source Heat Pumps. DOI: http://dx.doi.org/10.1016/B978-0-12-804220-5.00007-2

two-region vertical U-tube GHE after an experiment performed in a solar geothermal multifunctional HP experimental system. Lee et al. [18] conducted experiments on the thermal performance of a GCHP integrated into a building foundation in the summer. Man et al. [19] performed an *in situ* operational performance test of a GCHP system for cooling and heating in a temperate zone. The experimental results indicate that the performance of a GCHP system is affected by its intermittent or continuous operation modes. Petit and Meyer [20] compared the thermal performances of a GCHP with an air-source air conditioner (A/C), finding that a horizontal or vertical GCHP was more favourable in terms of economic feasibility. Esen and Inalli [21] proposed using the *in situ* thermal response test to determine the thermal properties of the ground for GCHP applications in Turkey, and they found that the thermal conductivity and effective thermal resistance of the ground vary slightly with depth.

The present chapter focuses on the energy and environmental analysis and modelling of a geothermal experimental plant from a continental temperate climate, located in an institutional building at the Polytechnic University of Timisoara, Romania. The system consists of a reversible vertical GCHP. One of the main innovative contributions of this study is in the achievement and implementation of an energy-operational optimisation device for the GCHP system using quantitative adjustment with a buffer tank (BT) and a variable speed circulating pump. Experimental measurements are used to test the performance of the GCHP system at different operating modes. The main performance parameters (energy efficiency and CO_2 emissions) are obtained for one month of operation using both classical and optimised adjustment of the GCHP system. A comparative analysis of these performances for both heating and cooling and DHW with different operation modes is performed. Additionally, two simulation models of thermal energy consumption in heating/cooling and DHW operation were developed using TRNSYS software. The simulations obtained using TRNSYS software are analysed and compared to the experimental measurements. The second objective of this chapter is to present the COP of a horizontal GCHP system and the temperature distributions measured in the ground heating season at University of Firat, Elazig, Turkey. Finally, a numerical model of heat transfer in the ground is described for determining the temperature distribution in the vicinity of the pipes.

7.2 PERFORMANCE OF A VERTICAL GCHP SYSTEM FOR HEATING, COOLING AND DHW OPERATION

7.2.1 Description of Experimental Laboratory

Experimental investigations of GCHP performance were conducted in a laboratory (Figure 7.1) at the Polytechnic University of Timisoara, Romania, located at the ground floor of the Civil Engineering Faculty building with six floors and a heated basement. The city has a continental temperature climate with four different seasons. The heating season in Timisoara runs from 1 October to 30 April, and the cooling season runs from 1 May to 30 September.

The laboratory room has an area of 47 m^2, and its height is 3.70 m. The envelope (external walls) is made of 200 mm of porous brick with a 100-mm thermal insulating layer and 20 mm of lime mortar. The thermal transmittances (U-values) are as follows: walls, 0.345 W/m^2K, and double-glazed windows, 2.22 W/m^2K. The area of the windows is 16 m^2, and the area of the interior door is 2.1 m^2. The indoor air design

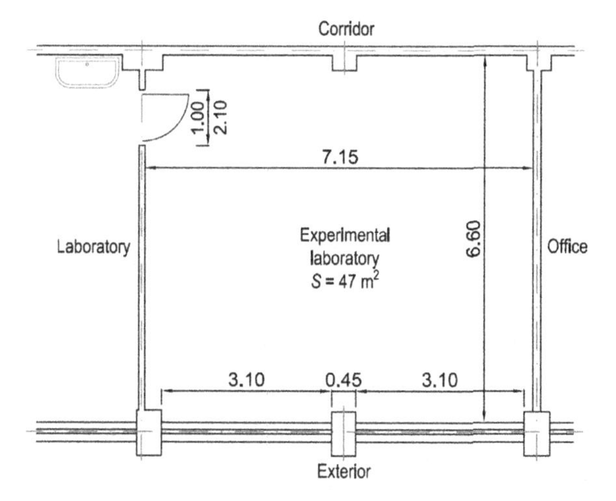

Figure 7.1 Experimental laboratory.

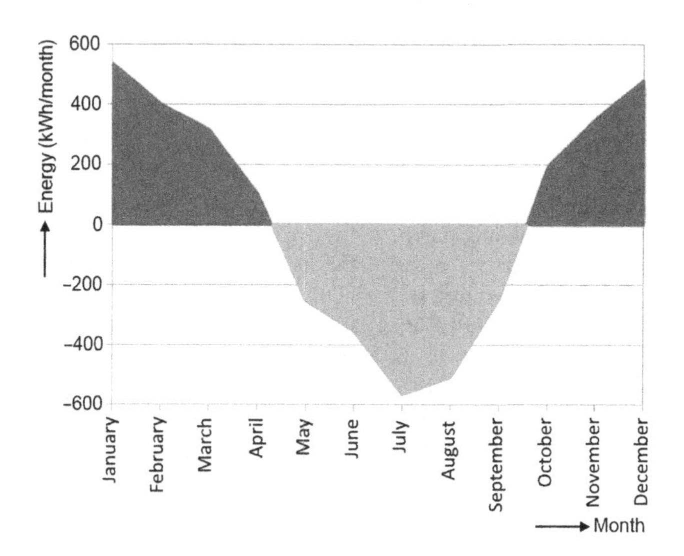

Figure 7.2 Monthly energy demand for laboratory heating/cooling.

temperature is 20 °C for the heating season and 26 °C for cooling season. The outdoor air design temperature is −15 °C for the heating season and 32.6 °C for the cooling season.

The GCHP installed in this experimental laboratory heated and cooled through a fan coil system. With the mentioned input data, a heating load of 3.11 kW and a cooling load of 2.15 kW were obtained. The laboratory area was assimilated with a three-person apartment area in Timisoara. Considering the DHW daily mean consumption of 50 l/person, a tank hot-water temperature of 45 °C and a cold water temperature of 20 °C, a DHW load of 4.36 kW was determined. Figure 7.2 illustrates the monthly energy demand for laboratory heating (positive values) and cooling (negative values).

7.2.2 Description of the Experimental System

The GCHP experimental system consisted of a BHE, HP unit, BT, circulating water pumps, fan coil units, sink, data acquisition instruments and auxiliary parts, as shown in Figure 7.3.

The heat carrier fluid can be delivered towards two fan coils units in two flow rate adjustment modes:

1. direct, by a recirculation pump connected inside of the HP unit of the GCHP system (classical solution);
2. indirect, by a fixed-speed circulating pump connected to a BT. The GCHP automation can control the operation of the circulating pump connected to the BT by on/off switching. This assembly improves the entire system operation. The BT allows decreasing the GCHP on/off switching because of its thermal inertia, and thus, the energy efficiency increases. The solution for heat carrier fluid flow rate adjustment was optimised using an automatic control device of circulating pump speed [22]. The main components of an automatic device for pump speed control are shown in Ref. [23].

Figure 7.4 presents a schematic of the automatic control device of the circulating pump speed according to heating/cooling demand of the room. In comparison with the classical solution, in which the circulating pump on/off switching is controlled by the GCHP automation, the optimised solution assures both the on/off switching and the speed control of the circulating pump.

The temperature difference between the inside and outside of the heated/cooled space is measured by temperature sensors TS1 and TS2 connected to a programmable logic controller (PLC). In the PLC internal memory, a computational algorithm of the frequency converter dependent on the measured temperature difference is implemented. The PLC sends the frequency converter the corresponding frequency value to ensure the fluid flow rate according to heating/cooling load at that time. In addition, the PLC allows on/off switching control of the circulating pump. For simultaneous on/off switching of the circulating pump and GCHP, common temperature values of this process were established.

7.2.2.1 Borehole Heat Exchanger

The GHE of this experimental GCHP consisted of a simple vertical borehole that had a depth of 80 m and was described in Chapter 6.

7.2.2.2 HP Unit

The HP unit was a reversible ground-to-water scroll hermetic compressor unit with R410A as a refrigerant. The nominal heating and cooling capacities were 6.5 (35 °C supply/0 °C return) and 3.8 kW (23 °C return/15 °C supply), respectively. The HP unit was a compact type model having an inside refrigeration system and DHW tank with a 175-l capacity. The operation of the HP was governed by an electronic controller, which, depending on the building water return temperature, switched the HP

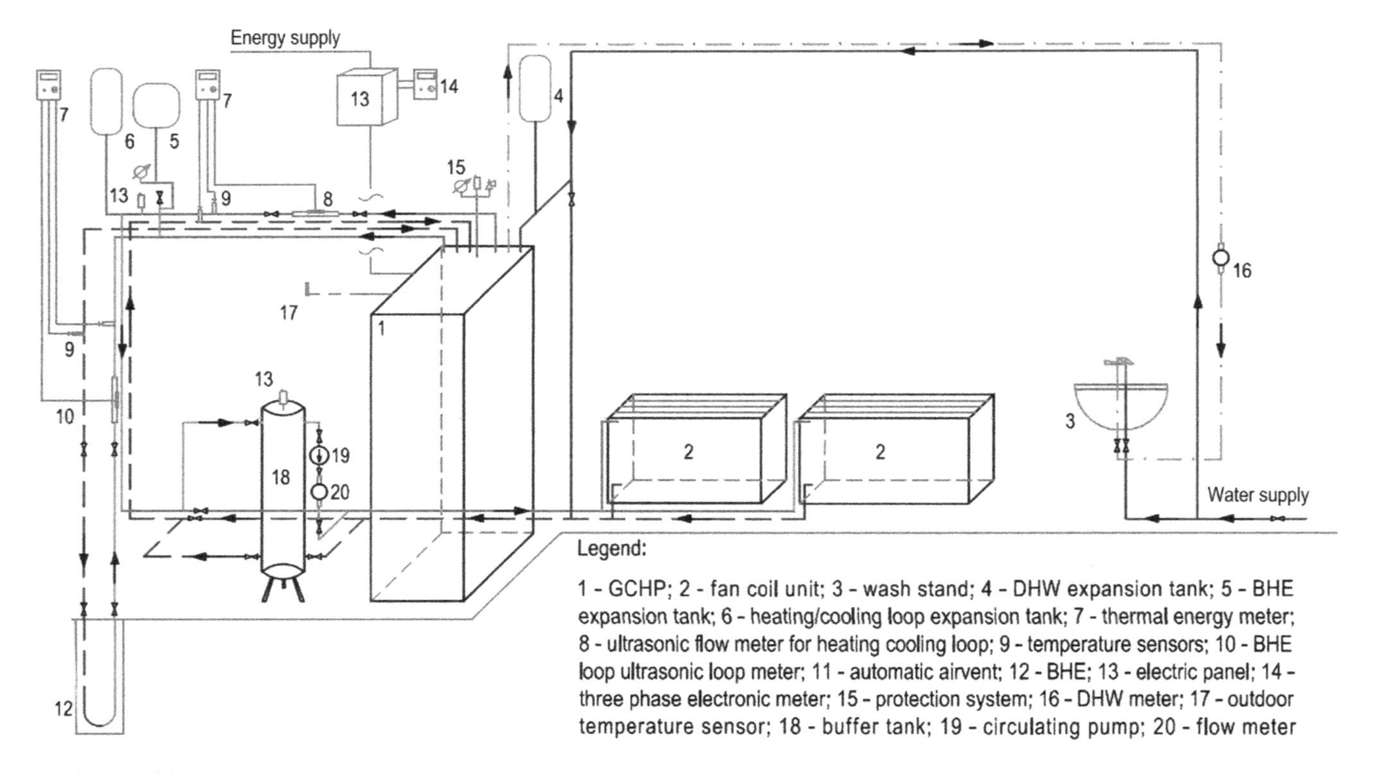

Figure 7.3 Schematic of the experimental GCHP system with optimised flow rate adjustment.

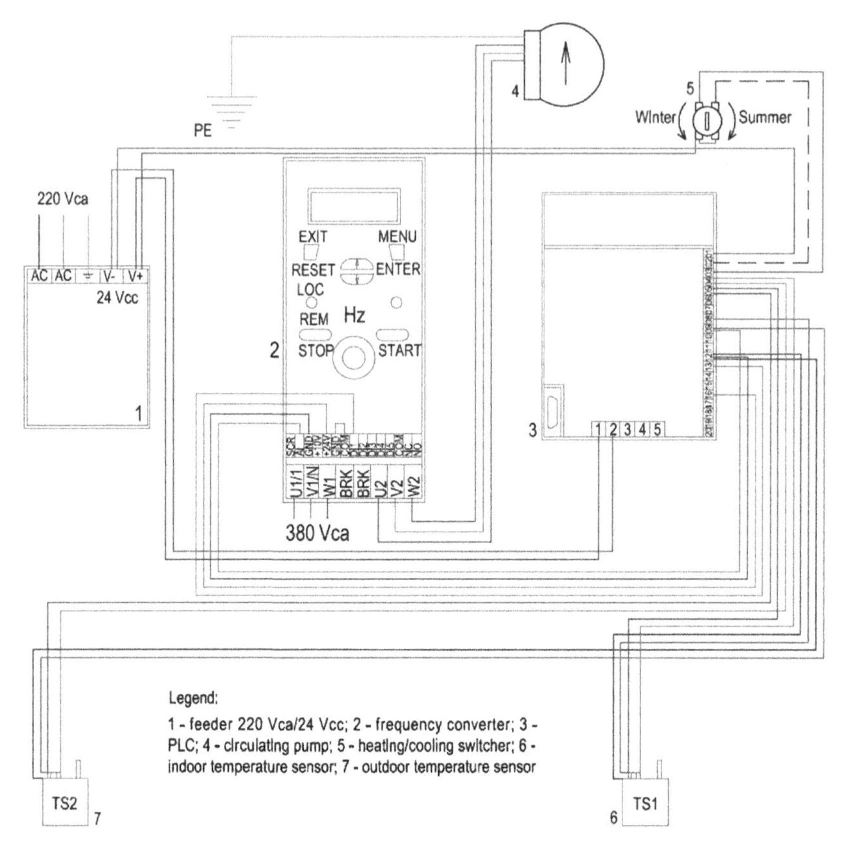

Figure 7.4 Schematic of the automatic control device for the circulating pump.

compressor on or off. The heat source circulation pump was controlled by the HP controller, which activates the source pump 30 s before compressor activation.

The COP of the GCHP system (COP_{sys}) is defined by Eqn (2.12), where E_{el} is the energy consumption of the GCHP system, which includes the energy consumption of the compressor of HP unit, circulating pumps, fan coil units, frequency converter and PLC.

The CO_2 emission (C_{CO_2}) of the heating system during its operation is calculated with Eqn (2.49).

To obtain the COP_{hp} or COP_{sys} and CO_2 emission, it is necessary to measure the heating/cooling energy E_t and electricity E_{el} used by the HP unit or the GCHP system.

7.2.2.3 Circulating Water Pumps

The water circulating loops of the GCHP consisted of a GCHP-BT water loop and BT-fan coil unit water loop. Two centrifugal pumps with rated flow of 2.8 and 5.5 m^3/h were chosen for the first and the second water circulating loop, respectively. The first circulating pump (fixed-speed circulating pump connected to the HP unit) was controlled by the GCHP automation, and the second pump (variable speed circulating pump connected to BT) was controlled by an automatic control device.

7.2.2.4 Fan Coil Units

Two parallel connected fan coil units were utilised as terminal units of the GCHP. The total thermal power of these two fan coil units was 3.2 kW.

7.2.2.5 GCHP Data Acquisition System

The GCHP data acquisition system consists of the indoor and outdoor air temperature, supply/return temperature, heat source temperature (outlet BHE temperature), DHW temperature, relative air humidity and main operating parameters of the system components.

7.2.3 Measuring Apparatus

Two thermal energy meters were used to measure the thermal energy produced by the GCHP and the extracted/injected thermal energy to the ground. A thermal energy meter was built with a heat computer, two PT500 temperature sensors and an ultrasonic mass flow meter. The two PT500 wires temperature sensors with an accuracy of $\pm 0.15\,^{\circ}\mathrm{C}$ were used to measure the supply and return temperature for a hydraulic circuit (the water-antifreeze solution circuit or the fan coil circuit). Also, an ultrasonic mass flow meter was used to measure the mass flow rate for a hydraulic circuit. The thermal energy meters were AEM meters, model LUXTERM, with a signal converter IP 67 and accuracy $<0.2\%$. A three-phase electronic electricity meter measured the electrical energy consumed by the system (the HP unit, the circulating pumps, a feeder 220 Vca/24 Vcc, a frequency converter and a PLC) and another three-phase electronic electricity meter measured the electrical energy consumed by the HP compressor. The two three-phase electronic electricity meters were a multifunctional type from AEM, model ENERLUX-T, with an accuracy grade in $\pm 0.4\%$ of the nominal value. The monitoring and recording of the experiments were performed using a personal computer. The indoor and outdoor air temperature was measured by air flow sensors and supply/return, heat source and DHW temperature was recorded by positive temperature coefficient immersion sensors, all connected to the GCHP data acquisition system and having an accuracy of $\pm 0.2\,^{\circ}\mathrm{C}$.

7.2.4 Laboratory Experiment Results

The system was monitored for 2 years. The experimental measurements were performed for two cases of flow rate adjustment of the heat carrier fluid in the system: case (1) – classical adjustment, by fixed-speed circulating pump connected to HP unit and case (2) – optimised adjustment, by variable-speed circulating pump connected to BT.

In heating operation, the experiments were conducted for a 1-month period for each of the two analysed cases, from the 22nd of January 2012 to the 20th of February 2012, and the 15th of January 2013 to the 13th of February 2013. The outdoor temperature varied in the range of -5.9 to $10.1\,^{\circ}\mathrm{C}$. The monthly mean values of the outdoor temperature during the two periods were almost equal. In cooling operation, the experiments were conducted for a 1-month period for each of the two analysed cases, from 21 May 2012 to 19 May 2012 and from 28 May 2013 to 26 June 2013. The outdoor temperature varied in the range of $15.2–34.8\,^{\circ}\mathrm{C}$. The monthly mean values of the outdoor temperature during the two periods were almost equal.

7.2.4.1 GCHP System Performances in Different Operation Modes

Heating operation. The experimental parameters including indoor air temperature (t_i) and outdoor air temperature (t_e), recorded for a 1-month period, are plotted in Figure 7.5. It should be noted that in case (2) a reduced indoor air temperature was obtained, around the set-point temperature of 22 °C, which led to better comfort in the room. In addition, the reduced oscillation of the antifreeze fluid temperature leads to a lower heat source demand. The heat source temperature in winter is up to 12–13 °C higher than the outdoor air temperature, this increases the capacity and the efficiency of a GCHP system.

Table 7.1 presents the summary of the mean values of the temperatures (t_i, t_e, t_s), electricity consumption (E_{el}), useful thermal energy for heating (E_t), COP of the GCHP system (COP_{sys}) and the HP unit (COP_{hp}), and CO_2 emission (C_{CO_2}).

The COP_{hp} values of the HP unit for the classical and optimised solutions are 4.82 and 5.06, respectively. For case (2), the $COP_{sys} = 4.40$ is 7.5% higher and the CO_2 emission level is 7% lower than in case (1). Due to the properties of the climate, ground, etc. of the place where the measurements were conducted and a higher underground water flow rate, the heat source temperature is increased and the COP_{hp} and COP_{sys} are notable values for both solutions. Therefore when these results were compared with results of the similar studies reported in Refs. [8,24] but in other local geothermal conditions, it is seen that the performance values obtained here are improved noticeably.

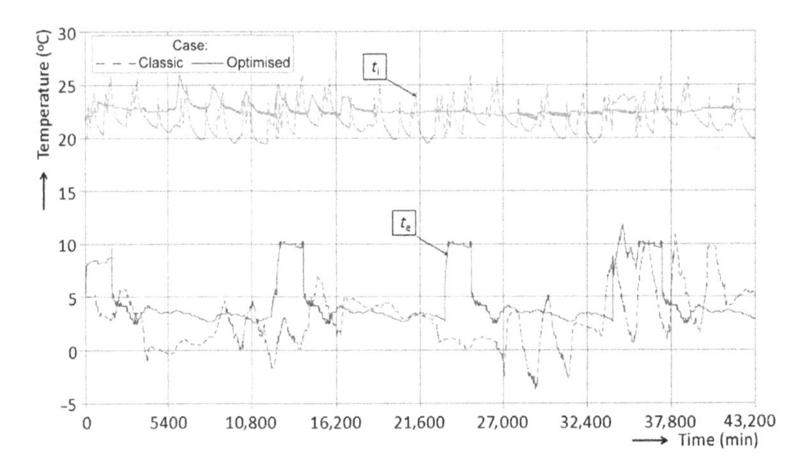

Figure 7.5 Recorded indoor and outdoor air temperature during heating operation.

Table 7.1 GCHP System Performance for Classical and Optimised Adjustment in Heating Operation

Case	t_i (°C)	t_e (°C)	t_s (°C)	E_{el} (kWh)	E_t (kWh)	COP_{sys}	COP_{hp}	C_{CO_2} (kg)
(1) Classic	22.65	3.25	16.24	125.18	510.62	4.07	4.82	50.45
(2) Optimised	21.84	3.76	17.08	116.47	512.54	4.40	5.06	46.94

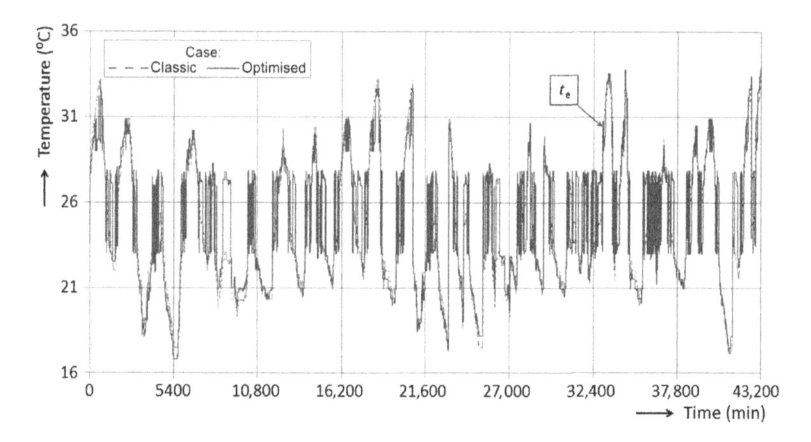

Figure 7.6 Outdoor air temperature evolution during the cooling provision tests.

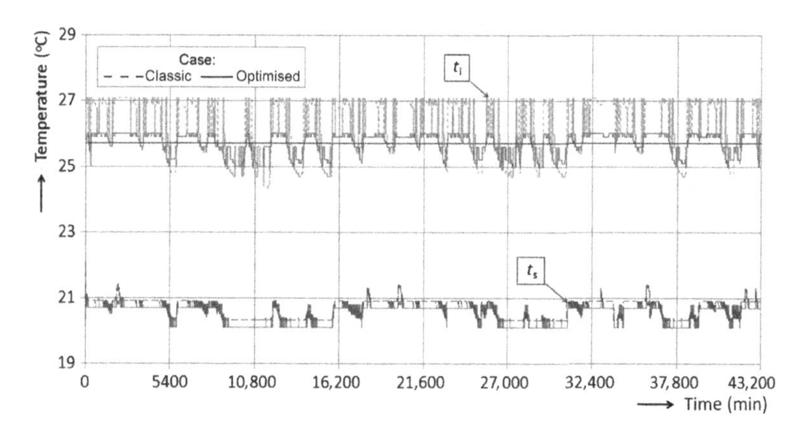

Figure 7.7 Recorded indoor air and heat source temperature during cooling operation.

Table 7.2 GCHP System Performance for Classical and Optimised Adjustment in Cooling Operation

Case	t_i (°C)	t_e (°C)	t_s (°C)	E_{el} (kWh)	E_t (kWh)	EER_{sys} (Btu/Wh)	COP_{sys}	C_{CO_2} (kg)
(1) Classic	26.22	24.54	20.50	70.99	287.21	13.79	4.04	28.60
(2) Optimised	25.71	24.88	20.65	65.16	288.45	15.09	4.42	26.25

Cooling operation. Figures 7.6 and 7.7 show the variation in time of the outdoor air temperature (t_e), indoor air temperature (t_i) and heat source temperature (t_s). In solution (2), a more reduced indoor air temperature around a set-point temperature of 26 °C was obtained, leading to better comfort in the room. Table 7.2 summarises the mean values

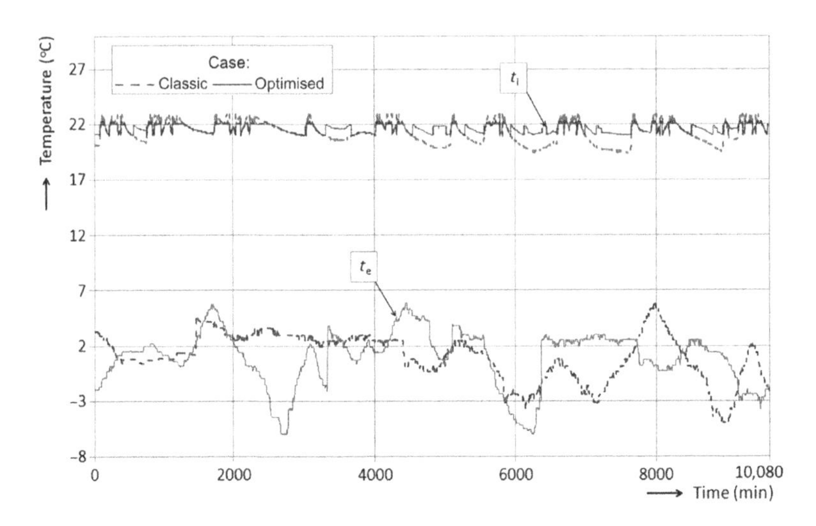

Figure 7.8 Recorded indoor and outdoor air temperature during heating and DHW operation.

of the temperatures (t_i, t_e, t_s), electricity consumption (E_{el}), useful thermal energy for cooling (E_t), EER of GCHP system (EER_{sys}) and CO_2 emission (C_{CO_2}). In case (2), the EER_{sys} was 8% higher, and the CO_2 emission level was 8% lower than in case (1).

The comparison of the GCHP system experimental performances in the heating and cooling operation (Tables 7.1 and 7.2) indicate that the system performance in heating and cooling operation was almost equal. This is because the heating load was higher than the cooling load, and in addition, the electricity consumption in heating operation was higher than the electricity consumption in cooling operation.

The COP values of the GCHP system were compared to the existing COP values applied in GCHP research. The research of Man et al. [19] revealed a COP for GCHP systems of 4.19–4.57 in the winter season and 3.9–4.53 in the summer season. In addition, the summer research of Michopoulos et al. [13], which used a vertical heat exchanger at a depth of 80 m, reported a COP_{sys} of 4.4–4.5. It is seen that the performance values obtained here are similar.

Heating and DHW operation. For the comparative study of the two flow-rate adjustment cases, the same DHW volume was used, $V_{dhw} = 1.22$ m^3. The experimental parameters including indoor air temperature (t_i), outdoor air temperature (t_e), DHW temperature (t_{dhw}) and heat source temperature (t_s), recorded for a week period for each of the two analysed cases, from 12th of January 2012 to 18th of January 2012, and 7th of January 2013 to 13th of January 2013, are plotted in Figures 7.8 and 7.9. The heat source temperature in winter is approximately 15 °C higher than the outdoor air temperature.

Table 7.3 presents the summary of the mean values of the temperatures (t_i, t_e, t_{dhw}, t_s), DHW volume (V_{dhw}), electricity consumption (E_{el}), useful thermal energy (E_t), COP of the GCHP system (COP_{sys}) and the HP unit (COP_{hp}) and CO_2 emission (C_{CO_2}).

Although the COP_{sys} resulted in almost equal values for the two cases, the experimental results indicate that, when using an automatic control device for the circulating pump speed, an electricity savings of 3% and a CO_2 emission level decrease of 3% were obtained.

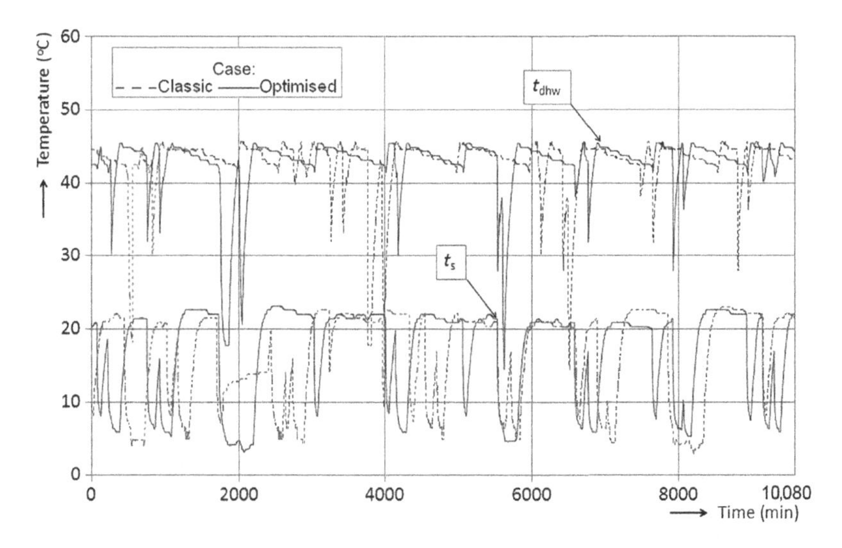

Figure 7.9 DHW and heat source temperature evolution during heating and DHW provision tests.

Table 7.3 GCHP System Performance for Heating and DHW Provision Tests

Case	t_i (°C)	t_e (°C)	t_s (°C)	t_{dhw} (°C)	V_{dhw} (m^3)	E_{el} (kWh)	E_t (kWh)	COP_{sys}	COP_{hp}	C_{CO_2} (kg)
(1) Classic	21.27	1.03	16.79	42.63	1.22	82.66	266.99	3.23	3.81	33.31
(2) Optimised	21.61	0.93	16.27	42.64	1.22	80.12	269.13	3.35	3.95	32.28

Analysing the experimental data (Tables 7.1 and 7.3) results indicate that the COP_{sys} of the GCHP system operating in heating and DHW mode in comparison with heating operation mode decreased significantly in the range of 20.6–23.9% in comparison to the two cases, from 4.07–4.40 to 3.23–3.35, respectively. The COP_{hp} values of the HP unit for cases (1) and (2) are 3.81 and 3.95, respectively.

Cooling and DHW operation. To determine the GCHP system performance in the summer season, experimental measurements over a 1-week period for each of the two analysed cases were performed, from 27 June 2012 to 3 July 2012, and 24 June 2013 to 30 June 2013. During the measurements, both the cooling and DHW load for a family using a DHW volume $V_{dhw} = 1.36\ m^3$ were assured. Figure 7.10 illustrates the evolution of indoor air temperature (t_i) and outdoor air temperature (t_e) and Figure 7.11 illustrates the evolution of the DHW temperature (t_{dhw}) and heat source temperature (t_s).

Table 7.4 presents a summary of the mean values of the temperatures (t_i, t_e, t_{dhw}, t_s), DHW volume (V_{dhw}), electricity consumption (E_{el}), useful thermal energy (E_t), COP of the GCHP system (COP_{sys}) and CO_2 emission (C_{CO_2}).

Although the COP_{sys} was almost equal for the two cases, the experimental results indicate that when using the automatic control device for the circulating pump speed, an electricity savings of 5% and a CO_2 emission level decrease of 5% were obtained.

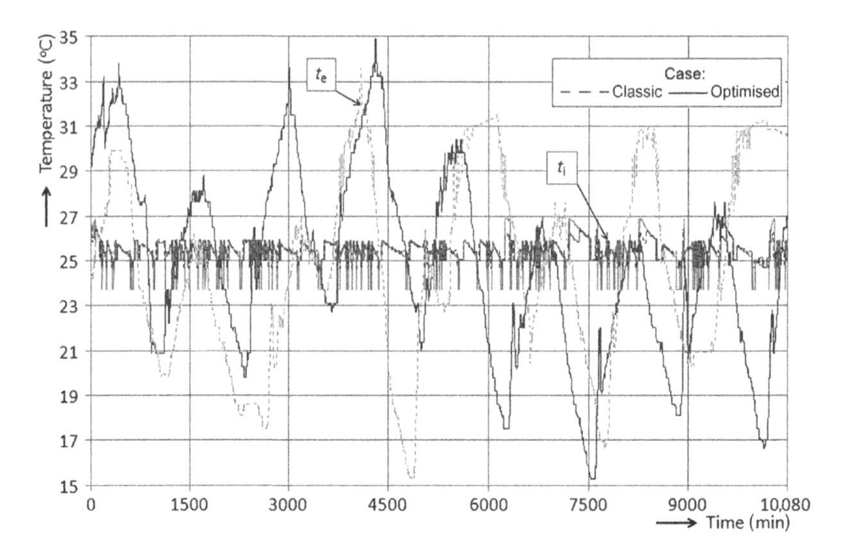

Figure 7.10 Recorded indoor and outdoor air temperature during cooling and DHW operation.

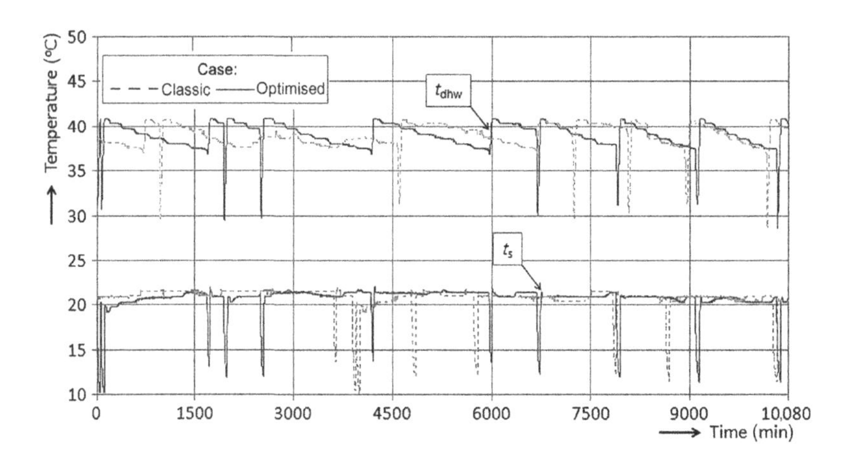

Figure 7.11 DHW and heat source temperature evolution during cooling and DHW provision tests.

Table 7.4 GCHP System Performance for Cooling and DHW Provision Tests

Case	t_i (°C)	t_e (°C)	t_s (°C)	t_{dhw} (°C)	V_{dhw} (m^3)	E_{el} (kWh)	E_t (kWh)	COP_{sys}	C_{CO_2} (kg)
(1) Classic	25.40	24.96	20.62	38.92	1.36	50.80	198.62	3.91	20.47
(2) Optimised	25.45	25.28	20.60	39.12	1.36	48.22	195,06	4.04	19.43

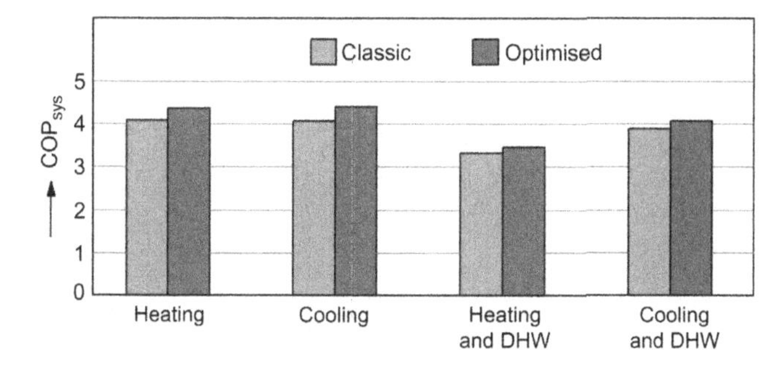

Figure 7.12 Variation of COP_{sys} in different operation modes of a GCHP system.

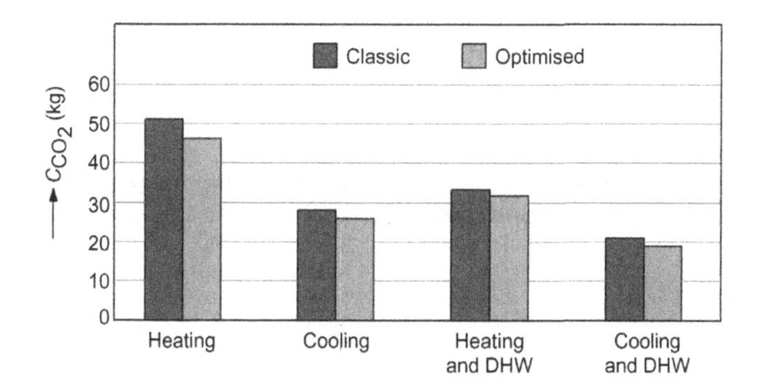

Figure 7.13 Variation of CO_2 emission in different operation modes of a GCHP system.

The experimental data (Tables 7.2 and 7.4) results of the COP_{sys} of the GCHP system operating in cooling and DHW mode in comparison with cooling operation mode indicate that there was a decrease only in the range of 3.2–8.6% compared with the two cases, from 4.04–4.42 to 3.91–4.04, respectively.

Figures 7.12 and 7.13 summarise the performances of the GCHP system in the different operation modes to show the experimental measurement results of the COP_{sys} and CO_2 emission (C_{CO_2}).

7.2.4.2 GCHP Performance in DHW Operation

The DHW production with different temperatures. To analyse the HP unit performances of the GCHP system that produces the DHW for a three-person family, a mean daily consumption of approximately 50 l/person at a DHW set-point temperature ($t_{dhw\text{-}set}$) of 40, 45, 50 and 60 °C was considered. The experimental measurements were conducted for a one-week period for each DHW set-point temperature from 1 April 2013 to 28 April 2013. The experimental parameters recorded for a DHW set-point temperature of 60 °C, including flow temperature (t_f), return temperature (t_r), DHW temperature (t_{dhw}) and heat source temperature (t_s), are plotted in Figure 7.14.

Table 7.5 presents the summary of the mean values of the temperatures (t_{dhw}, t_s), DHW volume (V_{dhw}), electricity consumption (E_{el}), useful thermal energy (E_t), COP

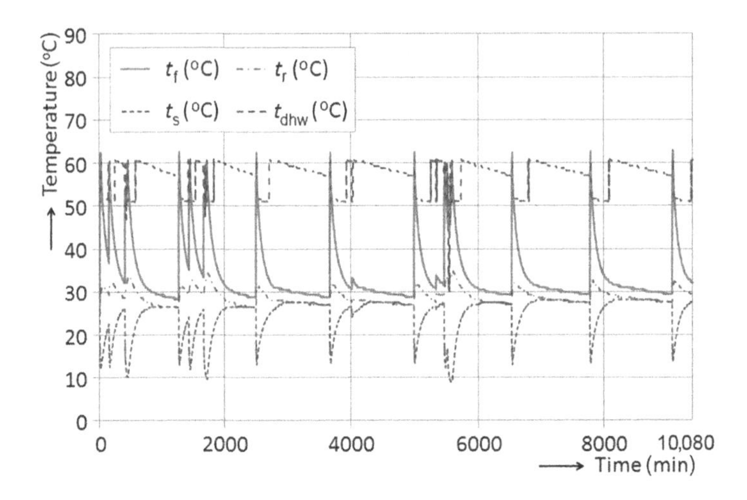

Figure 7.14 Recorded operation parameters of the GCHP during DHW provision tests for DHW set-point temperature of 60 °C.

Table 7.5 GCHP Performance During DHW Provision Tests

No.	$t_{dhw\text{-}set}$ (°C)	t_{dhw} (°C)	t_s (°C)	V_{dhw} (m^3)	E_{el} (kWh)	E_t (kWh)	COP_{hp}	C_{CO_2} (kg)
1	40	39.60	15.76	0.968	15.18	31.29	2.06	6.11
2	45	44.55	14.25	0.968	18.31	35.90	1.96	6.82
3	50	49.39	14.95	0.968	22.94	42.68	1.86	9.24
4	60	59.47	15.20	0.968	32.67	52.78	1.61	13.16

Table 7.6 GCHP Performance Depending on Water Temperature Increment in DHW Tank

No.	Δt (°C)	t_t (°C)	t_s (°C)	t_{dhw} (°C)	E_{el} (kWh)	E_t (kWh)	COP_{hp}	C_{CO_2} (kg)
1	3	34.10	12.78	17.79	0.170	0.556	3.33	0.068
2	5	40.29	13.83	29.49	0.310	1.011	3.26	0.124
3	10	37.13	10.46	26.10	0.720	2.264	3.14	0.290
4	15	38.66	11.53	26.76	0.980	2.952	3.01	0.394
5	20	42.34	9.53	30.01	1.440	4.044	2.81	0.580
6	25	43.72	10.12	31.50	2.010	5.128	2.55	0.810

of the HP unit (COP_{hp}) and CO_2 emission (C_{CO_2}). It should be noted that the HP's COP_{hp}, when operated in DHW mode, decreased from 2.06 to 1.61 if the DHW temperature increased from 40 to 60 °C, respectively.

Influence of water temperature increment in the DHW tank. The performance of the HP unit of the GCHP system is influenced by instantaneous consumed hot-water volume, which influences CO_2 emissions. Table 7.6 presents the summary of the mean

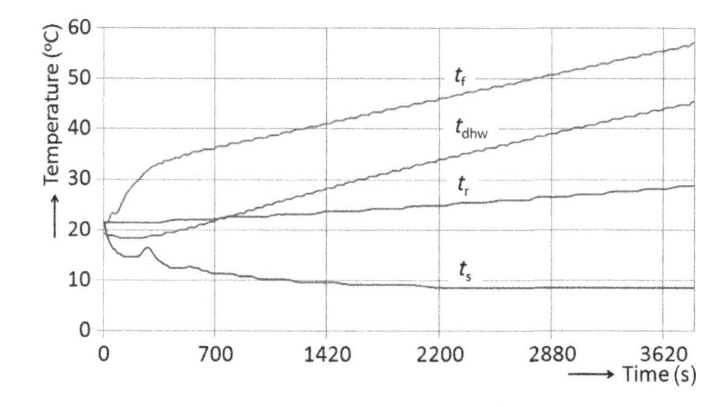

Figure 7.15 Recorded operation parameters of GCHP for water heating in a DHW tank with 25 °C.

values of the measured temperatures (t_f, t_r, t_{dhw}, t_s), electricity consumption (E_{el}), useful thermal energy (E_t), COP of the HP unit (COP_{hp}) and CO_2 emission (C_{CO_2}) for six experiments at different water temperature increments (Δt) in the DHW tank.

Figure 7.15 illustrates the evolution of heat carrier fluid temperatures (t_f, t_r) produced by GCHP, DHW temperature (t_{dhw}) and heat source temperature (t_s) for water heating in a DHW tank with $\Delta t = 25$ °C. The data in Table 7.6 indicates that with higher instantaneous DHW consumption (increased Δt), the COP_{hp} decreased. This COP_{hp} decrease could reach a level of 23% when the water temperature increment in the DHW tank is 25 °C.

7.2.4.3 Uncertainty Analysis

Uncertainty analysis (the analysis of uncertainties in experimental measurement and results) is necessary to evaluate the experimental data [25,26]. An uncertainty analysis was performed using the method described by Holman [25].

In the present study, the temperatures, thermal energy and electrical energy were measured with appropriate instruments explained previously. Error analysis for estimating the maximum uncertainty in the experimental results was performed using Eqn (6.43). It was found that the maximum uncertainty in the results is in the COP_{sys}, with an acceptable uncertainty range of 1.31–1.69% in heating operation mode and of 2.29–3.38% in cooling operation mode. The uncertainty of the COP_{hp} was estimated between 1.82% and 2.33% in heating mode and between 2.90% and 5.60% in DHW mode.

7.2.5 Numerical Simulation of Useful Thermal Energy Using TRNSYS Software

The basic principle of the TRNSYS program is the implementation of algebraic and first-order ordinary differential equations describing physical components into software subroutines (called types) with a standard interface. STEC library is based on steady state energy conservation formulated in thermodynamic quantities (temperature, pressure and enthalpy). In order to consider transient effects like start-up, a 'capacity' model has been developed that can be linked to each of the previously mentioned components and easily tuned to match empirical data. This permits thermal capacity to be considered only in those components where it has a large impact and avoids the huge gain in computational complexity of a full transient model. One of the main advantages of TRNSYS for the modelling and design of ground source HPs is that it includes components for the calculation of building thermal loads, specific components for heating/cooling, ventilating and A/C, HPs and circulating pumps, modules for BHEs and thermal storage, as well as climatic data files, which make it a very suitable tool to model a complete A/C HP installation to provide heating and cooling to a building.

Some statistical methods, such as the root-mean squared (RMS), the coefficient of variation (c_v), the coefficient of multiple determinations (R^2) and percentage difference (e_r) may be used to compare simulated and actual values for model validation.

The simulation error was estimated by the RMS, c_v, R^2 and e_r values computed using Eqns (6.20)–(6.22) and (6.44), respectively.

7.2.5.1 Simulation of Thermal Energy Used for Laboratory Heating and Cooling

Definition of the operation scheme. To simulate the thermal energy used to cover the heating/cooling load of the experimental laboratory, the operational connections were established between the building and all internal and external factors.

Figure 7.16 presents the operational scheme built in TRNSYS, where the building thermal behaviour was modelled using a 'Type 56' subroutine. This subroutine was

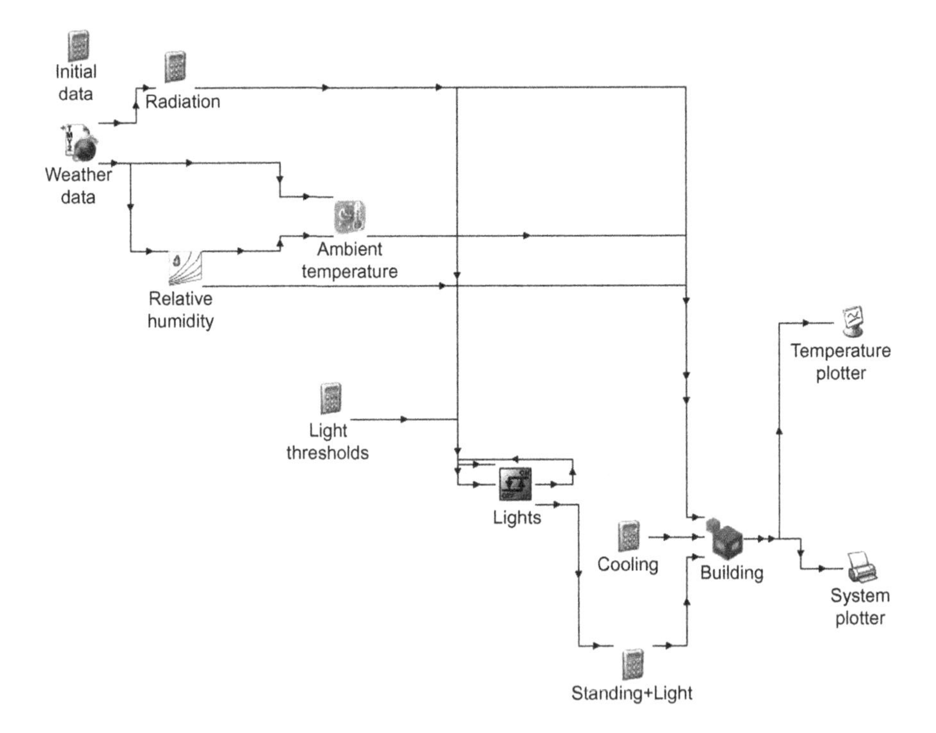

Figure 7.16 Scheme of the system model built in TRNSYS to simulate thermal energy consumption.

Table 7.7 Thermal Energy Used for Laboratory Heating and Cooling

Month	Heating Energy (kWh)		Percentage Difference e_r (%)	Cooling Energy (kWh)		Percentage Difference e_r (%)
	Simulated	Measured		Simulated	Measured	
January	505.00	512.48	1.45	0.00	0.00	0.00
February	391.40	390.12	0.32	0.00	0.00	0.00
March	303.21	300.87	0.77	0.00	0.00	0.00
April	99.45	97.89	1.59	0.00	0.00	0.00
May	0.00	0.00	0.00	208.00	211.23	1.52
June	0.00	0.00	0.00	242.00	242.82	0.33
July	0.00	0.00	0.00	337.90	333.12	1.43
August	0.00	0.00	0.00	437.80	445.14	1.64
September	0.00	0.00	0.00	324.30	319.20	1.59
October	189.70	191.31	0.84	0.00	0.00	0.00
November	348.90	345.23	1.06	0.00	0.00	0.00
December	477.50	480.21	0.56	0.00	0.00	0.00

Table 7.8 Statistical Values of Useful Thermal Energy of a GCHP System

Operation mode	RMS	c_v	R^2
Heating	2.72187	0.01409	0.99990075
Cooling	3.08003	0.02382	0.99977802
DHW production	8.50000	0.00464	0.99997906

processed with the TRNBuild interface by introducing the main construction elements, their orientation and surface, shadow factors and indoor activity type. Weather data for the Timisoara were obtained from the Meteonorm data base [27] and the weather data reader 'Type 109' and 'Type 89d' were used to convert the data in a form readable from TRNSYS. The simulation model took into account the outdoor air infiltration, heat/cold source type and interior gains. Also, light thresholds, cooling, shading and light integrators were defined for a good approach to the model. To extract the results, an online plotter ('Type 65') is used.

Simulation results and comparison with experimental data. Performing simulations for a 1-year period (8760 h), thermal energy used for heating and cooling were obtained and are presented beside the measured values in Table 7.7. Statistical values such as *RMS*, c_v and R^2 are given in Table 7.8 for the GCHP system in different operation modes. There was a maximum percentage difference between the TRNSYS simulated and measured values for the heating period of approximately 1.59% and for the cooling period of approximately 1.64%, which is very acceptable. The RMS and c_v values in heating mode are 2.722 and 0.0141, respectively and in cooling mode are 3.080 and 0.0238, respectively. The R^2-values in the two operation modes are about 0.9998, which can be considered as very satisfactory. Thus, the simulation model was validated by the experimental data.

7.2.5.2 Simulation of Thermal Energy Used to Produce DHW

Definition of the operation scheme. For simulation of the DHW production the operational scheme built in TRNSYS from Figure 7.17 was utilised.

The assembly of GCHP system consists of the standard TRNSYS weather data readers 'Type 15-6', a GCHP model 'Type 919', a BHE 'Type 557a' and a DHW storage tank 'Type 4', with a capacity of 175 l. Also, in the simulation model were defined single speed circulating pumps 'Type 114' for the antifreeze fluid in the BHE and 'Type 3d' for heat carrier fluid of the DHW coil. A 'Type 14' for the load profile and a daily load subroutine were created, this approach improving significantly the numerical convergence of the model. Finally, two model integrators ('Type 25' and 'Type 24') were used to calculate daily and total results for thermal energy produced.

Simulation results and comparison with experimental data. Useful thermal energy simulations for the assurance of the DHW thermal load were performed for four hot-water temperatures: 40, 45, 50 and 60 °C. The results of the simulation program are presented beside the experimental data in Table 7.9. Statistical values such as RMS and c_v are given in Table 7.8. A comparative analysis of these results indicates that the thermal energy values for DHW production simulated with TRNSYS were only 0.21–0.62% lower than the measured values in all four cases. The COP_{hp} values are in the range 1.56–2.00, approached to measured values (Table 7.5). The R^2-value of approximately 0.9999 is very satisfactory and the maximum percentage difference of 0.62% is acceptable and thus the simulation model is validated experimentally.

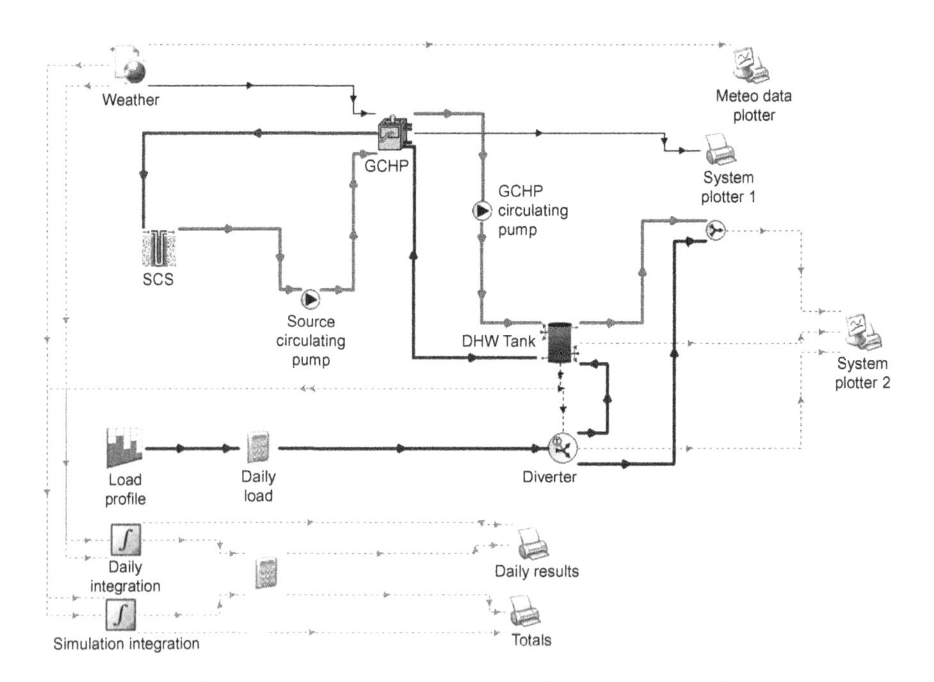

Figure 7.17 Scheme of the system model built in TRNSYS for simulation of the DHW production.

Table 7.9 Thermal Energy E_t Used for DHW Production

Temperature $t_{dhw\text{-}set}$ (°C)	E_t (kWh/year)		Percentage Difference e_r (%)	COP_{hp}
	Simulated	Measured		
40	1446	1449	0.21	2.00
45	1601	1611	0.62	1.90
50	1907	1913	0.31	1.80
60	2347	2359	0.51	1.56

7.2.6 Conclusions

The use of HPs in modern buildings with improved thermal insulation and reduced thermal load is a good alternative to classical heating/cooling and DHW solutions. It is recommended whenever possible to use terminal units capable of covering both heating and cooling demand. Indoor A/C can be a product of these coupled processes. This chapter has presented the evaluation of the performances of a GCHP system providing heating/cooling and DHW to an experimental laboratory. Some main conclusions can be deduced from this study:

1. The performed experimental research demonstrated higher performances of the GCHP system for the flow rate adjustment case using a BT and an automatic control device for circulating pump speed versus a classical adjustment case (COP_{sys} 7–8% higher and 7.5–8% lower CO_2 emission level).
2. The GCHP system, operating in heating or cooling mode, had a $COP_{sys} > 4$, and the GCHP system operating in heating or cooling and DHW mode had a $3 < COP_{sys} < 4$ for both cases.
3. In classical and optimised adjustment cases, the COP_{hp} values for heating and DHW provision tests were 3.81 and 3.95, respectively, and for heating operation tests, they were 4.82 and 5.06, respectively.
4. When using the circulating pump speed control, electricity savings and a reduction of the CO_2 emission of 3% for laboratory heating and 5% for laboratory cooling were obtained at the same time as DHW production.
5. If the GCHP is used to produce only DHW for a family at different temperatures between 40 and 60 °C, then the COP_{hp} would decrease to approximately 2, and the CO_2 emission level would vary between 6.11 and 13.16 kg.
6. For an instantaneously consumed hot-water volume, the energy performance of the GCHP can be decreased by up to 23% when the hot-water temperature in the DHW tank must be increased to 25 °C.
7. The developed TRNSYS simulation models can be used as a tool to determine the GCHP performance in different operation modes to optimise the system energy efficiency and ensure the user's comfort throughout the year.

7.3 NUMERICAL AND EXPERIMENTAL ANALYSIS OF A HORIZONTAL GCHP SYSTEM

The temperature distributions in the ground are very important for calculating the heat losses of buildings to the ground, for design of thermal energy storage equipment and

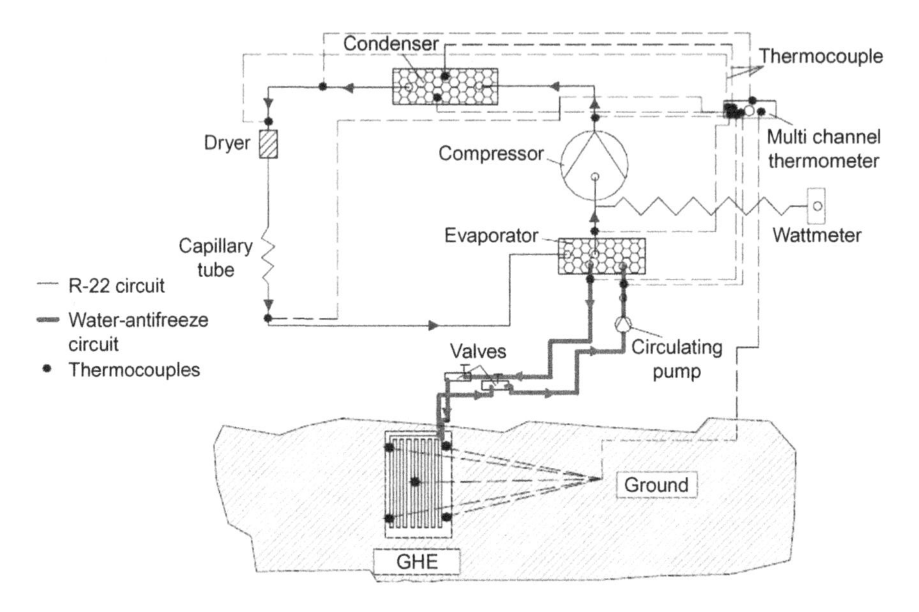

Figure 7.18 Schematic diagram of the experimental apparatus.

GHEs, and for analysis of the biodegradation processes of organic substances. Esen et al. [9] developed a numerical model of heat transfer in the ground for determining the temperature distribution in the vicinity of a horizontal ground heat exchanger (HGHE). The finite difference approximation is used for numerical analysis. In the experimental study, the COP_{sys} of the GCHP system and temperature distributions measured in the ground during the 2002–2003 heating season was presented.

7.3.1 Experimental Facilities

The schematic of the horizontal GCHP system constructed for space heating [9] is illustrated in Figure 7.18. Table 7.10 summarises the main component specifications and characteristics of the GCHP system. The experimental set-up consists of three main components: (1) the HGHE, (2) the HP unit and (3) the auxiliary equipment.

7.3.1.1 Horizontal Ground Heat Exchanger

One GHE has been installed at the University of Firat, Elazing, Turkey, which consists of a high-density polyethylene tube, 16 mm in diameter. This heat exchanger is made as a single pass straight tube, buried at a depth of 1 m. To allow for measuring the circulating water-antifreeze solution and ground temperature, a number of T-type thermocouples were installed.

The pipe-ground interface temperature was measured in a similar fashion to the water-antifreeze solution temperature measurement, except that thermal insulation is not used here because the thermocouple should have good contact with both the pipe and the ground. The ground temperature distribution around the pipe was measured by means of thermocouples distributed in the ground at various distances from the

Table 7.10 Technical Features of the Experimental Setup	
Specification	**Value/Characteristic**
Weather information (yearly average values)	
Average outdoor temperature	226 K
Maximum outdoor temperature	291 K
Minimum outdoor temperature	280 K
Average relative humidity	56%
Average solar radiation	14.9 MJ/(m^2day)
Average wind velocity	2.5 m/s
Average ground temperature at 1 m depth	289 K
Room information	
Window area	2.24 m^2
Wall area	34.63 m^2
Floor area	16.24 m^2
Ceiling area	16.24 m^2
Comfort temperature	293 K
Building volume	55.21 m^3
HP information	
Capacity	4.279 kW
Compressor type	Hermetic
Evaporator type	TT3; cooper and inner cooling aluminium
Condenser type	HS 10; Friterm
Compressor power input	2 HP; 1.4 kW
Compressor volumetric flow rate	7.6 m^3/h
Compressor rotation speed	2900 rot/min
Condenser fan	2350 m^3/h; 145 W
Evaporation temperature	0°C
Condensation temperature	54.5°C
Refrigerant type	R22
Ground heat exchanger information	
Configuration type	Horizontal
Pipe material	Polyethylene; PX-b cross link
Length of pipe	50 m
Pipe diameter	0.016 m
Piping depth	1 m
Pipe distance	0.3 m
Circulating pump information	
Type	Alarko; NPVO-26-P
Power	40, 62 and 83 W

pipes, in both the horizontal and vertical directions. Five thermocouples were used to measure the soil temperature around the HGHE (Figure 7.18).

7.3.1.2 Auxiliary Equipment

The collector valve varies the circulating water-antifreeze solution flow rate. The flow rate of the circulated water-antifreeze solution through the closed loop HGHE was measured using a rotameter and controlled by a hand-controlled tap mounted on the collector.

An anemometer was used to measure the circulating air flow velocity. The electric power consumed by the system (compressor, water-antifreeze circulating pump and fan) was measured by means of a wattmeter. The inlet and outlet temperatures of the R-22 in the condenser, compressor and evaporator were measured with T-type (copper-constantan) thermocouples. In addition, the temperatures of the circulated water-antifreeze solution at the inlet and outlet of the HGHE and evaporator (Figure 7.18) were measured. The ground temperature at 1 m depth was also measured next to the HGHE. The ambient and indoor air temperatures were measured with thermometers. The inlet and outlet pressures of the compressor and evaporator were measured using Bourdon tube-type manometers.

7.3.1.3 Heat Pump Unit

The heat transfer from the ground to the HP or from the HP to ground was maintained with the fluid or water-antifreeze solution circulated through the HGHE. The fluid transferred its heat to the refrigerant in the evaporator (the water-antifreeze solution to the refrigerant heat exchanger). The refrigerant, which flows through the other closed loop in the HP, evaporates by absorbing heat from the water-antifreeze solution circulated through the evaporator and then enters the hermetic compressor. The refrigerant is compressed by the compressor and then enters the condenser, where it condenses.

After the refrigerant leaves the condenser, the capillary tube provides almost 10°C of superheat that essentially gives a safety margin to reduce the risk of liquid droplets entering the compressor. A fan blows across the condenser to move the warmed air to the room. A non-toxic propylene glycol solution (25% weight) was circulated through the HGHE. In the heating season, the heat exchange fluid (water-antifreeze solution) in the HGHE loop collects heat from the ground and transfers that heat to the room. After the heat exchange fluid absorbs heat from the ground, the closed-loop HGHE circulates the heat exchange fluid through pipes buried 1 m deep in the trench (Figure 7.18).

7.3.2 Experimental Analysis and Uncertainty

The performance of the GCHP system is determined by measuring the flow rate and the temperature change of the fluid (water-antifreeze) and the electrical power input. The heat extracted by the unit in the heating mode (i.e. HGHE load) q_E is expressed as:

$$q_E = m_f c_f (t_{f,e} - t_{f,i}) \tag{7.1}$$

The electric power input to the compressor $P_{e,k}$, the water-antifreeze circulating pump $P_{e,p}$ and the condenser fan $P_{e,cf}$ can be written, respectively, as [9]:

$$P_{e,k} = I_k U_k \cos\varphi \tag{7.2}$$

$$P_{e,p} = I_p U_p \cos\varphi \tag{7.3}$$

$$P_{e,\mathrm{cf}} = I_{\mathrm{cf}} U_{\mathrm{cf}} \cos \varphi \tag{7.4}$$

The $\mathrm{COP}_{\mathrm{sys}}$ can be obtained based on Eqn (2.12) as:

$$\mathrm{COP}_{\mathrm{sys}} = \frac{Q_t}{P_{e,k} + P_{e,p} + P_{e,\mathrm{cf}}} \tag{7.5}$$

where Q_t is the useful thermal power supplied by the HP:

$$Q_t = m_{\mathrm{air}} c_{\mathrm{air}} (t_{\mathrm{air},o} - t_{\mathrm{air},i}) \tag{7.6}$$

The mean values of the measured data and the calculated results for December 2002 are given in Table 7.11. Using Eqn (7.5) and the values given in the table, the average $\mathrm{COP}_{\mathrm{sys}}$ value of the GCHP system was found to be 3.06 for December 2002.

Another important issue is the accuracy of the measured data as well as the obtained results by the experimental studies. Therefore, uncertainty analysis was

Table 7.11 Mean Values of the Measured Data and Calculated Results

Data	Value	Total Uncertainty (%)
Measured parameters		
Evaporation pressure	0.21 MPa	± 2.72
Condensation pressure	1.5 MPa	± 2.72
Evaporating temperature	261 K	± 2.75
Condensing temperature	313 K	± 1.38
Water-antifreeze temperature at HGHE inlet, $t_{f,i}$	280.3 K	± 1.38
Water-antifreeze temperature at HGHE outlet, $t_{f,e}$	283.6 K	± 2.89
Water-antifreeze mass flow rate, m_f	0.055 kg/s	± 1.38
Air mass flow rate, m_{air}	0.6 kg/s	± 3.00
Outdoor air temperature, t_{air}	290 K	± 3.00
Compressor electric current, I_k	5.98 A	± 3.00
Circulating pump electric current, I_p	0.32 A	± 3.00
Condenser fan electric current, I_{cf}	0.69 A	± 3.00
Current of all systems	6.99 A	± 3.00
Mono-phase voltage	220 V	± 1.38
Temperature of air at fan inlet, $t_{\mathrm{air,i}}$	295.8 K	± 1.38
Temperature of air at fan outlet, $t_{\mathrm{air,o}}$	303 K	± 1.00
Power factor, $\cos \varphi$	0.92	
Calculated parameters		
Power input to the compressor, $P_{e,k}$	1.211 kW	± 4.35
Power input to the circulating pump, $P_{e,p}$	0.065 kW	± 4.35
Power input to the condenser fan, $P_{e,cf}$	0.14 kW	± 4.35
Total power of systems, P_e	1.416 kW	± 4.35
Heat extraction rate from the depth of 1 m	0.671 kW	± 4.35

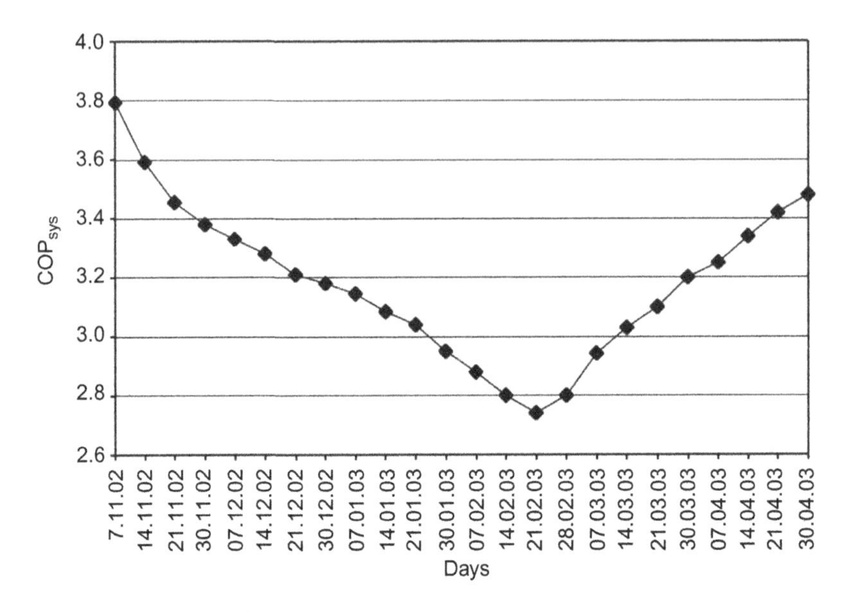

Figure 7.19 Daily performance of a GCHP system.

carried out to validate the experimental results. The total uncertainties of the other measured parameters calculated by Eqn (6.43) are presented in Table 7.11.

7.3.3 Daily Evaluation of the System

Figure 7.19 shows the daily periodic heating coefficient (COP_{sys}) of the GCHP system [9]. The daily periodic variation of COP_{sys} is given for a heating season lasting from 7 November 2002 to 30 April 2003. As seen from Figure 7.19, the system performance decreases until February, and then it increases. Because the lowest daily average temperature of the ground at 1 m depth is in February, the mean daily value of COP_{sys} was of 3.2. The highest COP_{sys} was found to be 3.8 (November), while the lowest COP_{sys} was 2.7 (February). The COP_{sys} increases with increasing the buried depth of the HGHE [8]. Obviously, a superior HGHE design and more careful ground selection will give higher enhancement rates for COP_{sys}.

7.3.4 Numerical Model for Ground Temperature Field

The natural ground temperature is a periodic function versus depth and time. Without considering moisture migration, the temperature field is a two-dimensional non-steady heat conduction problem with no internal heat generation and is not modelled with time-periodic boundary conditions. It is assumed that heat is transferred mainly by conduction, while convection and radiation contributions are negligible. The other assumptions made in the analysis are as follows:

1. the ground properties are uniform and the ground type does not change along the pipe;

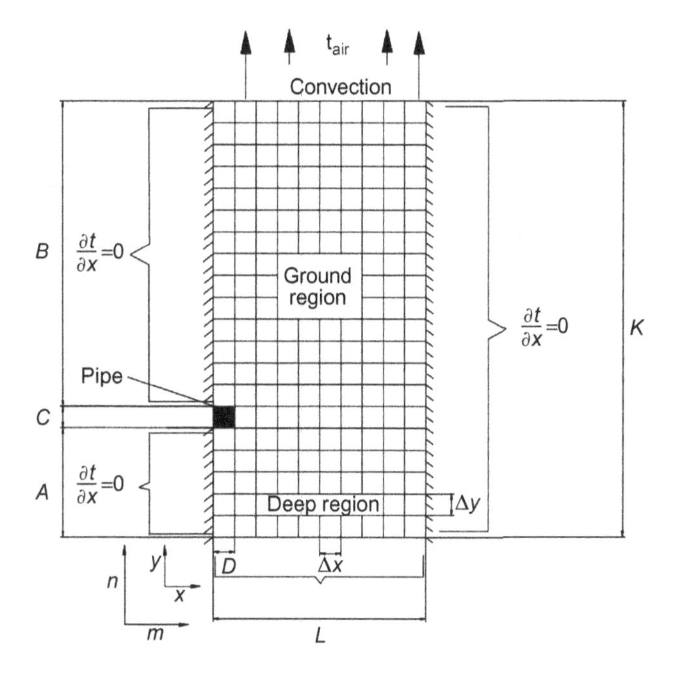

Figure 7.20 Finite-difference grid and boundary conditions of modelled domain.

2. the ground temperature at a certain distance from the pipe is assumed to change only with diurnal and seasonal variation and does not depend on the HGHE operation;
3. the heat transfer in the ground is assumed to be symmetric;
4. for the case of multiple HGHEs, the distance between loops is assumed to be large enough to avoid thermal interference between the loops;
5. the heat transfer in the ground in the direction parallel to the pipe is negligible;
6. the air-ground surface boundary is assumed to be convective;
7. the influence of gravity on the soil moisture transfer in the unsaturated soil is assumed to be negligible;
8. the rainfall or snow peak effect is assumed to be negligible.

The discretisation for this model is shown in Figure 7.20, while the meshes for different boundaries are shown in Figure 7.21.

The A, B, C, D, L and K letters are the distances in the x and y direction in the considered rectangular ground domain used for the calculation of the ground temperature distribution. Square meshes were considered in the underground field. In this case, the solution is easy and the solution time is short. The $x-y$ plane of the rectangular calculation domain was divided into nodal spaced Δx and Δy apart in the x and y directions, respectively. A general interior node (m, n) has the coordinates $x = n\Delta y$, as shown in Figure 7.20. In this study, an equal distance of mesh $(\Delta x = \Delta y)$ is selected, and distance in the z direction is considered as $\Delta z = 1$. The total lengths are 0.16 and 1.312 m in the x and y directions, respectively. The physical properties of this ground

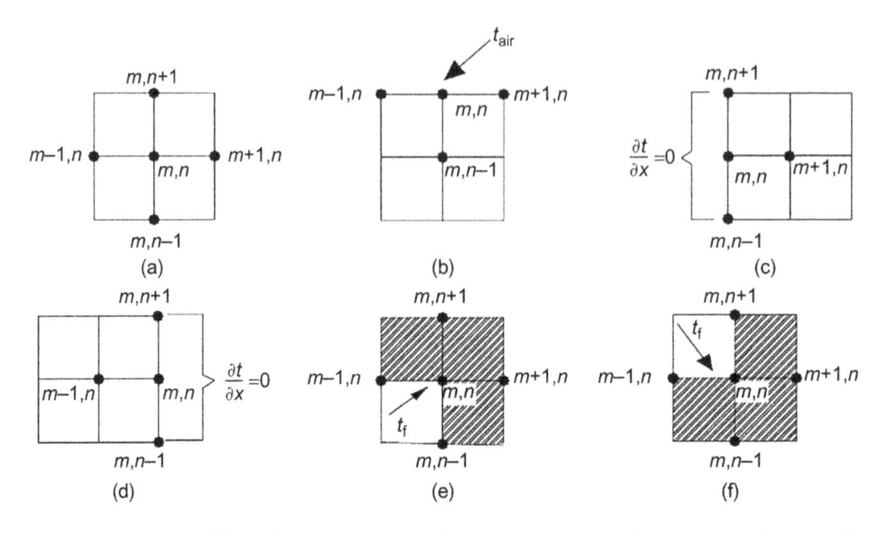

Figure 7.21 Meshes for different boundaries. (a) Conduction (inner volume); (b) Convection (upper surface of ground); (c) Symmetry (left-upper or left-bottom surface); (d) Symmetry (right surface); (e) Pipe surface; (f) Pipe surface (different).

are assumed to be constant for simulation and equal to the physical properties used in the experiments.

The two-dimensional transient heat conduction equation for the ground is represented by:

$$\frac{\partial^2 t(x,y,\tau)}{\partial x^2}+\frac{\partial^2 t(x,y,\tau)}{\partial y^2}=\frac{1}{a}\frac{\partial t(x,y,\tau)}{\partial \tau} \tag{7.7}$$

where $t(x,y,\tau)$ is the ground temperature, in °C; x is the horizontal distance, in m; z is the vertical distance, in m; a is the thermal diffusivity of ground, in m^2/s; τ is the time, in s.

The initial and boundary conditions are:

for $\tau = 0$:

$$t(x,y,\ 0)=t_d \tag{7.8}$$

for $0<y<A$ and $x=0$:

$$\frac{\partial t(x,y,\tau)}{\partial x}=0 \tag{7.9}$$

for $A<y<(A+C)$ and $x=0$:

$$t(x,y,\tau)=t_f(x,y,\tau)=\text{constant} \tag{7.10}$$

for $A+C<y<(A+C+B)$ and $x=0$:

$$\frac{\partial t(x,y,\tau)}{\partial x}=0 \tag{7.11}$$

for $0<y<(A+B+C)$ and $x=L$:

$$\frac{\partial t(x,y,\tau)}{\partial x}=0 \tag{7.12}$$

for $0 < x < L$ and $y = 0$:

$$t(x, y, \tau) = t_{\mathrm{d}} \tag{7.13}$$

for $0 < x < D$ and $y = A$:

$$t(x, y, \tau) = t_{\mathrm{f}}(x, y, \tau) \tag{7.14}$$

for $0 < x < D$ and $y = (A + C)$:

$$t(x, y, \tau) = t_{\mathrm{f}}(x, y, \tau) \tag{7.15}$$

for $0 < x < L$ and $y = (A + C + B)$:

$$-\lambda \frac{\partial t(x, y, \tau)}{\partial y} = k_{\mathrm{air}}[t(x, y, \tau) - t_{\mathrm{air}}] \tag{7.16}$$

The finite-difference procedure used is that first introduced by Barakat and Clark [28], and is an unconditionally stable explicit method.

The finite-difference equation (FDE) of Eqn (7.7) is [9]:

$$\frac{t^p_{m+1,n} + t^p_{m-1,n} - 2t^p_{m,n}}{(\Delta x)^2} + \frac{t^p_{m,n+1} + t^p_{m,n-1} - 2t^p_{m,n}}{(\Delta y)^2} = \frac{1}{a}\frac{t^{p+1}_{m,n} - t^p_{m,n}}{\Delta \tau} \tag{7.17}$$

where p is the time step, m is the node count in the x direction and n is the node count in the y direction.

The transient FDE for an interior node can be expressed on the basis of Figure 7.21a:

$$t^{p+1}_{m,n} = \mathrm{Fo}(t^p_{m-1,n} + t^p_{m+1,n} + t^p_{m,n+1} + t^p_{m,n-1}) + (1 - 4\mathrm{Fo})t^p_{m,n} \tag{7.18}$$

The coefficient of $t^p_{m,n}$ in the $t^{p+1}_{m,n}$ expression is $1-4\mathrm{Fo} \geq 0$, which is independent of the node number (m, n), and thus the stability criterion for all interior nodes in this case is $1-4\mathrm{Fo} \geq 0$:

$$\mathrm{Fo} = \frac{a \Delta t}{(\Delta x)^2} \leq \frac{1}{4} \tag{7.19}$$

where Fo is the mesh Fourier number; a is the thermal diffusivity of the ground, in m^2/s; Δt is the time step, in s; $\Delta x = \Delta y$ is the distance between two nodes.

The thermal diffusivity and the thermal conductivity for moist clay ground are $a = 6.71 \cdot 10^{-7}\ m^2/s$ and $\lambda = 2.2\ W/(m \cdot K)$, respectively. The distance between two nodes or diameter of pipe is $\Delta x = 0.016$ m. The time step Δt is calculated from the stability criterion given in Eqn (7.19) for inner nodes, and is found to be $\Delta t \leq 95.38$ s [9].

The transient FDE for an upper surface of ground area can be expressed on the basis of Figure 7.21b as:

$$t^{p+1}_{m,n} = 2\mathrm{Fo}\left[t^p_{m,n-1} + \frac{1}{2}(t^p_{m-1,n} + t^p_{m+1,n}) + \mathrm{Bi}t_{\mathrm{air}}\right] + (1 - 4\mathrm{Fo} - 2\mathrm{FoBi})t^p_{m,n} \tag{7.20}$$

The stability criterion is $\mathrm{Fo} \leq 1/[2(2 + \mathrm{Bi})]$ at the upper surface of ground, where Bi is the Biot number of the system. The heat transfer coefficient k_{air} is calculated by the following equation:

$$k_{\mathrm{air}} = 2.8 + v_{\mathrm{wind}} \tag{7.21}$$

where $v_{\mathrm{wind}} = 23$ m/s is the average wind velocity.

The thermal diffusivity and the thermal conductivity of air are determined with dry air table properties for an average outdoor air temperature of 7.3°C. The time step has been taken as $\Delta t = 1$ s for this simulation [9].

The transient FDE for an upper left surface (B) or a bottom left surface (A) of ground region can be expressed on the basis of Figure 7.21c as:

$$t_{m,n}^{p+1} = \mathrm{Fo}\left[\frac{2}{1}t_{m+1,n}^{p} + t_{m,n-1}^{p} + t_{m,n-1}^{p}\right] + (1 - 4\mathrm{Fo})t_{m,n}^{p} \quad (7.22)$$

The transient FDE for a right surface (K) of the ground region can be expressed on the basis of Figure 7.21d as:

$$t_{m,n}^{p+1} = \mathrm{Fo}\left[\frac{2}{1}t_{m-1,n}^{p} + t_{m,n-1}^{p} + t_{m,n+1}^{p}\right] + (1 - 4\mathrm{Fo})t_{m,n}^{p} \quad (7.23)$$

As the wall thickness of the pipe is very small (2 mm) the pipe conductive resistance can be neglected. The transient FDE for an inner surface of pipe can be expressed on the basis of Figure 7.21e and f as follows:

$$t_{m,n}^{p+1} = \frac{4}{3}\mathrm{Fo}\left[\frac{1}{2}(t_{m-1,n}^{p} + t_{m,n-1}^{p}) + t_{m,n+1}^{p} + t_{m+1,n}^{p} + \mathrm{Bi}t_{\mathrm{f}}^{p}\right] + \left(1 - 4\mathrm{Fo} - \frac{4}{3}\mathrm{FoBi}\right)t_{m,p}^{p} \quad (7.24)$$

$$t_{m,n}^{p+1} = \frac{4}{3}\mathrm{Fo}\left[\frac{1}{2}(t_{m-1,n}^{p} + t_{m,n+1}^{p}) + t_{m,n-1}^{p} + t_{m+1,n}^{p} + \mathrm{Bi}t_{\mathrm{f}}^{p}\right] + \left(1 - 4\mathrm{Fo} - \frac{4}{3}\mathrm{FoBi}\right)t_{m,n}^{p} \quad (7.25)$$

The criterion is found to be $\mathrm{Fo} \leq 3/[4(3 + \mathrm{Bi})]$ at the inner surface of pipe. The average velocities of water-antifreeze fluid v_{f} have been determined by the following equations:

$$v_{\mathrm{f}} = \frac{m_{\mathrm{f}}}{\rho_{\mathrm{f}} A_{\mathrm{pipe}}} \quad (7.26)$$

where m_{f} is the mass flow rate of water-antifreeze fluid, in kg/s; ρ_{f} is the density of water-antifreeze fluid; A_{pipe} is the cross-section area of the pipe. The input data for numerical study are outdoor air temperature, fluid temperature and mass flow rate of fluid. They were taken from the experimental measurements. The average temperature of water-antifreeze solution is 12.4°C. Density, kinematic viscosity, thermal conductivity and thermal diffusivity, corresponding to this temperature was determined.

7.3.5 Model Validation

The model results were compared to the experimental data. Temperatures were determined at every half hour experimentally and numerically. The ground temperatures were obtained from the numerical and experimental study for different mass flow rates of the water-antifreeze solution. The change of the ground temperature for a flow rate of 0.041 kg/s is shown in Figure 7.22 [9]. The ground temperature decreases with time because the amount of thermal energy within the ground decreases. In addition, the increasing mass flow rate results in decreasing ground temperature because the high mass flow rates extract more energy from the ground. The average difference between the numerical and experimental values is approximately 0.6°C. This value shows a good agreement between the numerical and experimental results.

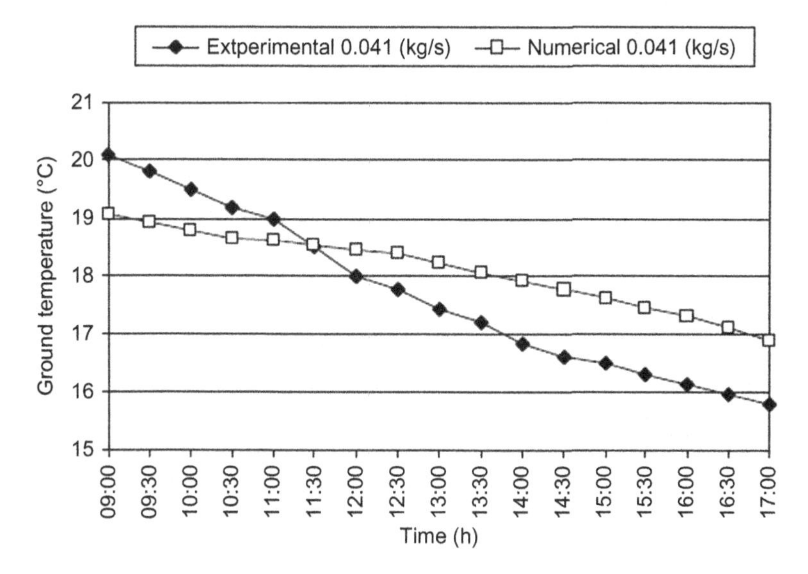

Figure 7.22 Ground temperature change for fluid mass flow rate as 0.041 kg/s.

7.3.6 Conclusions

The main conclusions that can be drawn from this study are listed below:

1. The average value of the COP_{sys} of the GCHP system was found to be 3.2 in the 2002–2003 heating season. Future modifications will significantly improve the performance of the system.
2. The numerical results showed good agreement with the experimental data.
3. The heating load of the GCHP depends on the ground temperature distribution around the HGHE. The temperature distribution is important to the performance enhancement of the GCHP and, especially for the HGHE.
4. The heating seasonal performance factor of the system with 1-m depth of the HGHE was found to be 2.12.
5. The relevant ground characteristics are to be precisely measured and controlled during GCHP operation before attempting the design of the HGHE. Therefore, care must be taken in the design and construction of a ground loop for an HP application to ensure long ground loop life and reduce the installation cost.

REFERENCES

[1] Yang W, Zhou J, Xu W, Zhang GQ. Current status of ground-source heat pumps in China. Energy Policy 2009;38(1):323–32.

[2] Yang H, Cui P, Fang Z. Vertical-borehole ground-couplet heat pumps: a review of models and systems. Appl Energy 2010;87:16–27.

[3] Bayer P, Saner D, Bolay S, Rybach I, Blum P. Greenhouse gas emission savings of ground source heat pump systems in Europe. Renew Sustainable Energy Rev 2012;16:1256–67.

[4] Sarbu I, Sebarchievici C. General review of ground-source heat pump system for heating and cooling of buildings. Energy Build 2014;70(2):441–54.

[5] Lund JW, Freeston DH, Tonya L, Boyd TL. Direct utilization of geothermal energy 2010 worldwide review. In: Proceedings of the world geothermal congress. Bali, Indonesia; 25–29 April, 2010. p. 1–23.
[6] Pahud D, Mattthey B. Comparison of the thermal performance of double U-pipe borehole heat exchanger measured in situ. Energy Build 2001;33(5):503–7.
[7] Bose JE, Smith MD, Spitler JD. Advances in ground source heat pump systems – an international overview. In: Proceedings of the 7th international conference on energy agency heat pump. Beijing, China; 2002. p. 313–324.
[8] Inalli M, Esen H. Experimental thermal performance evaluation of a horizontal ground-source heat pump system. Appl Therm Eng 2004;24(14-15):2219–32.
[9] Esen H, Inalli M, Esen M. Numerical and experimental analysis of a horizontal ground-coupled heat pump system. Build Environ 2007;42(3):1126–34.
[10] Yavuzturk C. Modelling of vertical ground loop heat exchangers for ground source heat pump systems [Doctoral thesis]. USA: Oklahoma State University; 1999.
[11] Retkowski W, Thoming J. Thermoeconomic optimization of vertical ground-source heat pump systems through nonlinear integer programming. Appl Energy 2014;114:492–503.
[12] Zeng HY, Diao NR, Fang ZH. Efficiency of vertical geothermal heat exchangers in ground source heat pump systems. Int J Therm Sci 2003;12(1):77–81.
[13] Michopoulos A, Bozis D, Kikidis P, Papakostas K, Kyriakis NA. Three-year operation experience of a ground source heat pump system in Northern Greece. Energy Build 2007;39(3):328–34.
[14] Mostafa H, Sharqawy SA, Said EM. First *in situ* determination of the ground thermal conductivity for borehole heat exchanger applications in Saudi Arabia. Renew Energy 2009;34(10):2218–23.
[15] Hwang YJ, Lee JK, Jeong YM, Koo KM, Lee DH, Kim LK, et al. Cooling performance of a vertical ground-coupled heat pump system installed in a school building. Renew Energy 2009;34:578–82.
[16] Pulat E, Coskun S, Unlu K. Experimental study of horizontal ground source heat pump performance for mild climate in Turkey. Energy 2009;34:1284–95.
[17] Yang WB, Shi MH, Liu GY. A two-region simulation model of vertical U-tube ground heat exchanger and its experimental verification. Appl Energy 2009;86:2005–12.
[18] Lee JU, Kim T, Leigh SB. Thermal performance analysis of a ground-coupled heat pump integrated with building foundation in summer. Energy Build 2013;59:37–43.
[19] Man Y, Yang H, Wang J, Fang Z. *In situ* operation performance test of ground couplet heat pump system for cooling and heating provision in temperate zone. Appl Energy 2012;97:913–20.
[20] Petit PJ, Meyer JP. Economic potential of vertical ground-source heat pumps compared to air-source air conditioners in South Africa. Energy 1998;23(2):137–43.
[21] Esen H, Inalli M. *In-situ* thermal response test for ground source heat pump system in Elazig, Turkey. Energy Build 2009;41:395–401.
[22] Sebarchievici C. Optimization of thermal systems from buildings to reduce energy consumption and CO_2 emissions using ground-coupled heat pump [Doctoral Thesis]. Romania: Polytechnic University Timisoara; 2013.
[23] Thornton JW, McDowell TP, Shonder JA, Hughes PJ, Pahud D, Hellstrom G. Residential vertical geothermal heat pump system models: calibration to data. ASHRAE Trans 1997;103(2):660–74.
[24] Esen H, Inalli M, Esen M. Technoeconomic appraisal of a ground source heat pump system for a heating season in eastern Turkey. Energy Convers Manage 2006;47(9-10):1281–97.
[25] Holman JP. Experimental method for engineers. Singapore: McGraw Hill; 2001.
[26] Hepbasli A, Akdemir O. Energy and exergy analysis of a ground source (geothermal) heat pump system. Energy Convers Manage 2004;45(5):737–53.
[27] METEONORM Help, Version 5.1; 2004.
[28] Barakat HZ, Clark JA. On the solution of the diffusion equations by numerical methods. ASME J Heat Transfer 1966;88:421–7.

INDEX

Note: Page numbers followed by "*f*" and "*t*" refer to figures and tables, respectively.

G

H

S

Printed in the United States
By Bookmasters